THE COMPLICATED ROOF

A Cut and Stack Workbook

Companion Guide to *A ROOF CUTTER'S SECRETS*

by Will Holladay

W&H PUBLISHERS

THE COMPLICATED ROOF
A Cut and Stack Workbook
Companion Guide to *A Roof Cutter's Secrets*

Copyright © 2009 by Will Holladay

Book design by	Jaime Vásquez
	Millennium Systems Corp.
	Panama, Rep. of Panama.
Cover production by	Bill Grote
	Craftsman Book Company
	Carlsbad, California
Cover photos by	
upper:	Fred Stockwell Photography
	Ashland, Oregon
lower:	Hawkeye Aerial Photography
	Santa Clara, California
Illustrations by	Will Holladay

Library of Congress Catalog Control Number: 2008911139
International Standard Book Number: 9780945186014

W&H Publishers
Lompoc, California

Printed in the United States of America

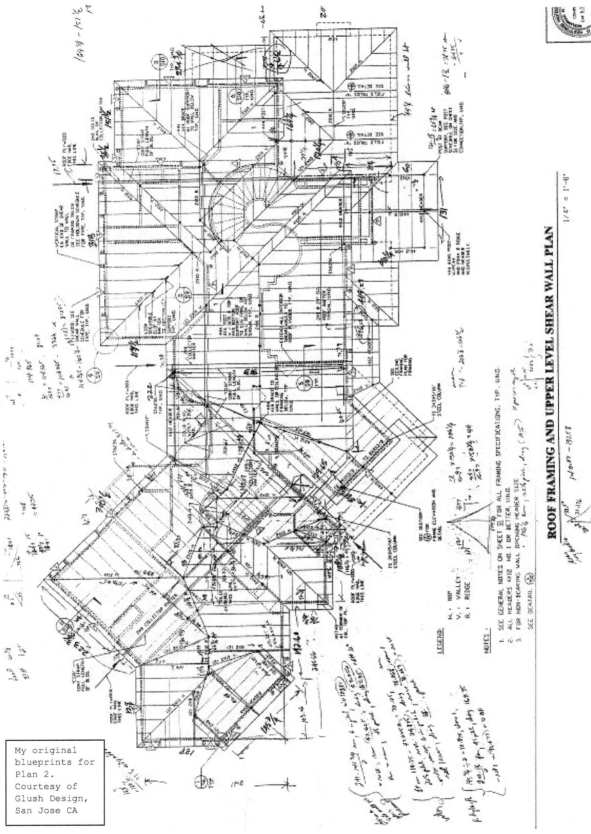

ROOF FRAMING AND UPPER LEVEL SHEAR WALL PLAN

1/4" = 1'-0"

LEGEND:
H. : HIP
V. : VALLEY
R. : RIDGE

NOTES:

1. SEE GENERAL NOTES ON SHEET S2 FOR ALL FRAMING SPECIFICATIONS, TYP. U.N.O.
2. ALL HEADERS 4X12 MIN. 1 OR BETTER, U.N.O.
3. FOR NON-BEARING WALL OPENING HEADER SIZE
   SEE DETAIL

My original
blueprints for
Plan 2.
Courtesy of
Glush Design,
San Jose CA

i

But those who hope in the Lord will renew their strength. They will soar on wings like eagles; they will run and not grow weary, they will walk and not be faint.
Isaiah 40:31

With a little help from my friends.

Thank you
Chuck Cline, Mike Diamond (Calculated Industries), Bill Grote (Craftman Book Company), Nick Ridge, Dave Saunders, Rick and Susan Tyrell, and Jaime Vasquez (Millennium Systems Corp.).

TABLE OF CONTENTS

Preface.. v

Abbreviations vi

Book Guidelines vii

Roof 1 – Medford OR 1997, 10/12 Pitch, JD Resner Construction

Questions 1-1, -2, -3, roof ratios and rafter heelstands.................................. 1

Questions 1-4, -5, -6, -9, common rafters... 1

Questions 1-7, -8, TOR heights 2

Questions 1-10, -11, -12, -13, -14, -20, regular hips 4

Questions 1-15, -16, -17, -18, -19, supported valleys 6

Question 1-21, dual purpose hip/valley 9

Questions 1-22, -23, -24, -25, ridge lengths 10

Question 1-26, shed dormer 11

Question 1-27, wall valley 12

Questions 1-28, -29, tall wall and rake wall 13

Questions 1-30, -31, -32, -33, -34, -35, -36, -37, -38, bay roofs 15

Question 1-39, special gable end commons 24

Questions 1-40, -41, -42, -43, regular hip and valley jacks 25

Questions 1-45, -46, regular parallel hip-valley jacks 27

Question 1-46, regular diverging hip-valley jacks 27

Question 1-47, over-under intersection 28

Roof 2 – Monte Sereno CA 2000, 6/12 Pitch, SR Freeman Construction

Phase 1

Questions 2-1, -2, -3, roof ratios and rafter heelstands................................... 31

Question 2-4, common rafters.. 32

Questions 2-5, -6, TOR heights 32

Question 2-7, regular hips and valleys 35

Questions 2-8, -9, ridge lengths 36

Questions 2-10, -11, -22, broken hips 36

Question 2-12, finding plate height difference 39

Questions 2-13, -14, regular hip and valley jacks 40

Questions 2-15, -16, -19, regular parallel hip-valley jacks 40

Questions 2-17, -18, -20, regular diverging hip-valley jacks 42

Question 2-21, california framed diverging hip-valley jacks 44

Question 2-23, roof skeleton schematic 49

Phase 2

Questions 2-24, -25, common rafters and TOR 51

Question 2-26, ridge lengths 53

Question 2-27, V6 and H9 55

Question 2-28, -32, dogleg hips and valleys 56

Question 2-29, V7 and H8 56

Question 2-30, -35, bastard broken hips 59

Question 2-31, H10 60

Question 2-33, DV3 and H11 61

Question 2-34, BH6 63

Question 2-36, -37, -38, -39, -40, dormer at R9 65

Questions 2-41, -42, -43, -44, -45, roof skeleton top connections 69

Question 2-46, regular hip and valley jacks 75

Question 2-47, regular parallel hip-valley jacks 75

Question 2-48, dogleg hip and valley jacks 75

Question 2-49, bastard broken hip jacks 76

Questions 2-50, -51, -52, -53, -54, -55, -56, -57, jack fill for non-standard areas 77

Cut-out copies of Roof Plans 93

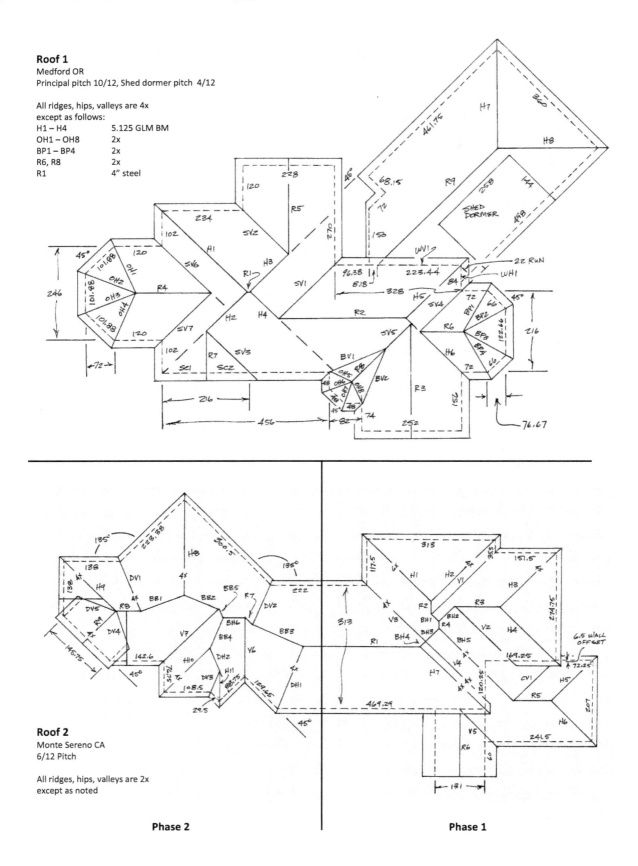

Roof 1
Medford OR
Principal pitch 10/12, Shed dormer pitch 4/12

All ridges, hips, valleys are 4x
except as follows:
H1 – H4   5.125 GLM BM
OH1 – OH8   2x
BP1 – BP4   2x
R6, R8   2x
R1   4" steel

Roof 2
Monte Sereno CA
6/12 Pitch

All ridges, hips, valleys are 2x
except as noted

Phase 2   **Phase 1**

PREFACE

This workbook is an application of the methods presented in *A Roof Cutter's Secrets* (RCS). By following alongside as I walk thru the process of solving two "real life" complicated roofs, one will have the best opportunity to learn how "this particular roof cutter" approaches and solves a difficult roof. This workbook parallels the thinking process that I use when solving these types of roofs. While there are many items included that would have normally only warranted a brief subconscious acknowledgement on my part, here in this workbook they are expanded upon by my choosing for the sake of the reader. As with any workbook where the answers are given directly following the question, the full benefit of the exercise can only be gotten when one uses a separate sheet of paper to hide the answer and associated drawing while challenging one's brain to answer the question without assistance. The method I used to solve any roof section is shown in the answer's explanation. If desired, a deeper study into a specific topic and it's methodology can be found by referring to RCS. As an aid to this process, the applicable section in RCS is noted along with the given answer/explanation. Explanations for a specific roof situation are only given once in the workbook. If an explanation was given for a situation on Roof 1 and the same or similar situation is found on Roof 2, only the calculations will be shown. Cut out copies of each of the roof plans are available at the back of the book. Having the applicable plan off to the side while answering the questions will save a lot of page flipping.

The two roofs featured in this workbook were also featured as the cover and interior cover of the 2002 revised edition of RCS. They were chosen in both cases because of the rich variety of roof framing situations found in each. I personally cut both roofs and helped lead the stacking on these projects. The house with the supporting hips and various bay roof extensions is located in Medford, Oregon and was framed by contractor Jack Resner in 1997. The house with the angular chopped up hip roof is located in Monte Sereno, California and was framed by contractor Shone Freeman in 2000. The roof for Shone is perhaps the most difficult roof I have cut in my lifetime. Both projects were great fun and involved a variety of very talented carpenters of whom I was privileged to be a one.

In a real job, after calculating any rafter length, I would write this measurement on the roof framing plan in its actual location (i.e.: hip, valley or common). I do the same for ridge heights and any calculated ridge lengths. I make small drawings around the borders of the plans for any special situations that need a little visual help to detail a connection or as reminders of how I decided to cut a particular section. Obviously, when one has all the necessary rafter lengths written on the plans, the puzzle has been solved and it is time to arrange the rafter material on racks to mark layout for gang cutting.

In addition to a copy of RCS you will need either a #4065 Construction Master® Pro V 3.1 series, #4080 CM Pro Trig or a #4050 CM5 (substitute M+ for Stor 1) to follow many of my answers. The workbook has been laid out so that one can solve the simpler questions with a regular calculator (without special roof functions), while the more difficult questions are solved using the CM calculator. This is the calculator I normally use and have used for many years. It made no sense to go thru all the complicated extra steps to solve the difficult questions with a regular calculator when the majority of you would be using a roof or trig calculator anyhow. The CM calculator can find angles and solve bastard rafters in no time. If you are using the CM calculator for all the workbook questions, you undoubtedly know how to enter the appropriate "roof pitch" into the calculators memory in place of applying the RR and LL ratios. You will save a few key strokes here and there. I often double check my results by using one method against the other (ratios vs CM calculator). Better yet, I like someone else to solve the same roof independently of any input by me to make sure I didn't make many errors. While there are bound to be mistakes in this book, I pray they are few and irrelevant. If you do spot one let us know.

Since this is a workbook on calculating "complicated roofs" NOT entry level roofs, little time is afforded for questions of that grade such as: the layout of rafters, birdsmouths, ridge-cuts, etc. Only in special case situations of the above might these types of questions be included. You will find a few wall related questions in the text. Typically, wall construction isn't considered part of roof cutting but does play an important role in roof stacking. You as a roof cutter will many times be asked for input on special situations like those included in the book.

Please note: I've used the "< >" symbols occasionally in equations to denote a subtotal or serve as benchmark of what should be in the calculator at that stage. Depending in what you last had in the calculator (roof pitch/12 or degrees) regulates the number of times "Pitch" must be pressed in the CM calculator to get the desired pitch or degrees.

I hope you enjoy the book and it helps to increase your roof framing skills. Most likely, this book will be my last. I thank the Lord once again for the opportunity to pass along what I have learned in this lifetime. God Bless you all.

ABBREVIATIONS

BB	bastard broken hip
BH	broken hip
BM	beam
BP	bastard hip
bldg	building
BV	bastard valley
calcs	calculations
Calif, CA	California or overlay style
CM	Construction Master®
COM LL ratio	common rafter line length ratio
CV	California valley
dbl	double
DH	dogleg hip
DV	dogleg valley
EF	effective
GLM	gluelam beam
H	regular hip
ht	height
H/V	hip/valley
H/V LL ratio	hip/valley line length ratio
LL	line length
LP	long point
OC	on-center
OH	octagon hips
RCS	A Roof Cutter's Secrets
RR ratio	roof rise ratio
rt	right
SC	special common
SP	short point
SV	supported valley
thk	thickness
TOR	top of ridge
Q	Question
V	regular valley
WV	wall valley
WH	wall hip
½ thk	one half the thickness of
½ 45° thk	one half the 45° thickness of
½ 22.5° thk	one half the 22.5° thickness of
< >	subtotal or benchmark in calculation

BOOK GUIDELINES

The book has been written in an "inches ONLY with decimals" format (no feet). This saves time calculating and I believe is less confusing. If you are unaccustomed to this format I suggest you spend a few minutes to memorize the 7 useful conversion values of decimals to fractions (.125 = ⅛, .25 =¼, .375 =⅜, .50 =½, .625 =⅝, .75 =¾, and .875 =⅞). By remembering that 1⁄16" is .06 you can adjust even smaller if so desired but rarely this is required. Rounding off to one of the above listed fractions is usually sufficient for framing purposes. You will also not see the little inch hash marks (") denoting inches in the answers and often not in the questions either. But since all measurements in the book are given in inches the (") marks are understood to be there even when not shown. The building dimensions noted will be considered as actual with exterior plywood sheathing installed. In other words, no adjustments need to be made - just use what is noted. I refer to particular walls by either calling them by their given length on the plans (ie: "wall 234") or by calling them out as between principal rafters (ie: "the wall between H1 and SV2"). Lumber size is regarded as ½" under nominal (ie: a 2x8 measures 1.5"x 7.5"). Rafter spacing will be considered as 24"OC. Common rafters have a standard seatcut dimension of 3".

If a question asks for "rafter length", it is referring to a shortened for ridge thickness line length (LL) dimension from headcut to heelcut measured along the bottom edge of the rafter. Most regular rafters are given in this form. Exceptions are noted as "theoretical LL" or "theoretical length" in the question, which refers to the length of a rafter to the top framing point without respect to any shortening at the top connection and/or the lower end (if applicable). The same holds true for ridges. Typically those types of questions are found in the more complex second phase of Roof plan 2. In that phase it is much easier to deal with theoretical lengths or theoretical heights and shorten during layout, after referring to a scale drawing of the non standard connection. Common rafters butting to a ridge board will be considered the standard ridge connection. Therefore, ridge height denotes top of ridge (TOR) above the outside wall height. Exceptions are called out ("ridge beam below", etc.). Right and left side jack rafter fill is referred to as if an individual was outside the building looking up the carrying rafter (hip or valley) from below. All rafter tails are run wild to be cut to length when hanging fascia.

Regarding regular Jack fill: In my method of cutting roofs only the "jack step length" is ever calculated and this dimension is recorded at the top of the roof plan page along with the RR/LL ratios and other constants. Actual jack lengths are only calculated in special situations (along a wall, etc). Since regular jack rafters either "step down" from a common rafter length or "step up" in multiples of the step length itself (depending on the situation), the jack step length and the common rafter lengths are all the information one needs when laying out to cut hip or valley fill. For this workbook consider the roof eaves to be open and the hip jacks step longer from the building corner (shortest hip jack to LP is the length of one jack step, 2nd longest jack is two steps to LP, etc.). This method gives consistency to the look of the eave from the corner and avoids the special hip corner frieze blocks since a flyer rafter will be installed where the edge of the hip leaves the building wall and all blocks are square-cut regular (RCS pg. 72). Consider valley jack rafters to step shorter from the common rafter length (RCS pg. 91).

If you are a stickler for carrying the OC spacing from a long straight run of common rafters out onto the fill, or always match layout across the ridge try to be open minded to the way these are done in the workbook. I do not waste time/effort calculating or cutting this way because I know the roof sheathers will handle any change in layout without even slowing down. They run the plywood sheets wild at the hips and must cut at the valley anyway (RCS pg. 141). Matching layout across the ridge is only beneficial when CJs serve as a tension tie. When the ridges are beams (Roof 1) or vertically braced (Roof 2) matching layout across the ridge serves little purpose (RCS pg. 93). All bastard pitches have been converted to an x/12 format so one can easily use a Quick Square® or Speed® Square for layout purposes. This also coincides with the Construction Master® calculator programming.

The words "span, run, effective run and travel" are important to understand. They are defined in RCS pgs. 1-3. Looking at a cross-section cut of the roof taken parallel with the common rafters, "span" is referred to as the total distance measured from the exterior face of an outside rafter wall to the exterior face of the opposing outside rafter wall. "Run" is half that distance and "effective run" is the horizontal distance covered by a common rafter from the outside wall to the near side of ridge. "Travel" is only used when referring to regular/irregular hips and valleys. It is the horizontal distance covered directly under one of these rafters from the outside wall to the ridge (in other words, the "noon-time shadow on the ground" of that rafter in position). "Theoretical travel" refers to a similar measurement taken to the center of the ridge.

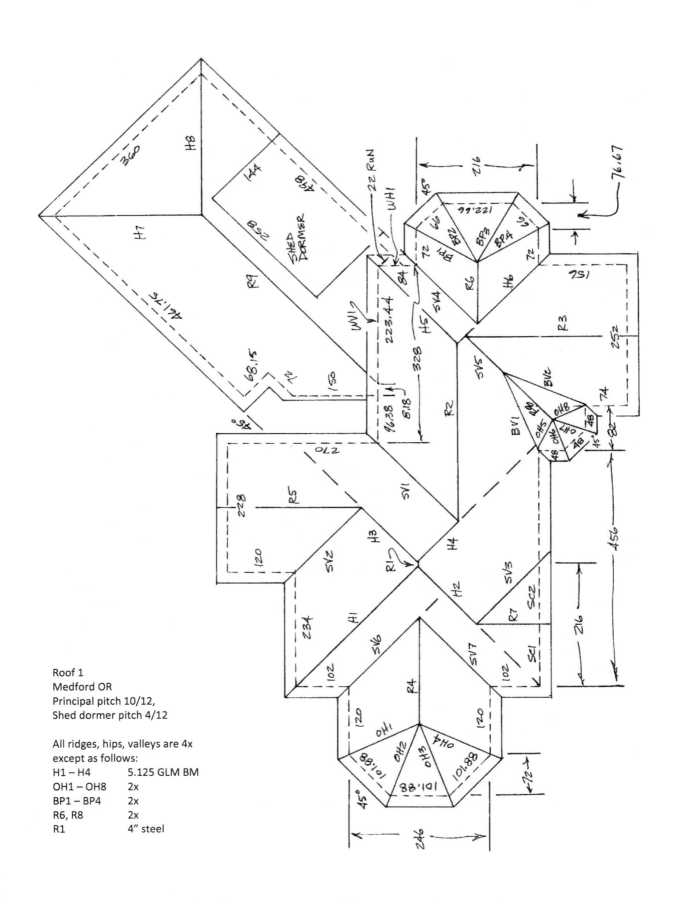

Roof 1
Medford OR
Principal pitch 10/12,
Shed dormer pitch 4/12

All ridges, hips, valleys are 4x
except as follows:
H1 – H4   5.125 GLM BM
OH1 – OH8   2x
BP1 – BP4   2x
R6, R8   2x
R1   4" steel

ROOF 1

The journey of 1000 miles begins with a single step.

Question 1-1

What is the RR ratio for this 10/12 pitch roof?

RCS pgs. 3, 299

Answer: **.8333**

        Math: $10 \div 12 = $ **.8333**

Question 1-2

What are the common rafter and hip/valley LL ratios for this 10/12 roof?

RCS pgs. 3, 4, 298

Answer: Com LL ratio = **1.3017**; H/V ratio = **1.6415**

        Math:

           SOLVED using CM calculator.

           10 Inch Pitch, 1 Run, Diag (**1.3017**), Hip/V (**1.6415**)

Question 1-3

Considering the roof pitch is 10/12, rafters are 2x8 and seatcut is 3", what is the heelstand?

RCS pgs. 1, 53

Answer: **7.26**

        Math:

           1) $3 \times .8333 = 2.5$

           2) $7.5 \times 1.3017, -2.5 = $ **7.26**

        Method: Calculate heelcut depth and subtract it from the plumb distance across 2x8 rafter.

Book note: The answers from questions 1-1 to 1-3 should be written at the top of the roof plan sheet to use as a reference.

Question 1-4

What are the lengths of the common rafters for R2, R3, R4 and R5 considering they all butt to 4x ridge beams?

RCS pgs. 49-50

Answers: 1) R2 = **192.98**

2) R3 = **161.74**

3) R4 = **157.83**

4) R5 = **146.12**

Math:

1) 300 -3.5, ÷2, x1.3017 =**192.98**
2) 252 -3.5, ÷2, x1.3017 =**161.74**
3) 246 -3.5, ÷2, x1.3017 =**157.83**
4) 228 -3.5, ÷2, x1.3017 =**146.12**

Method: Span less thk of ridge, divided by 2 and multiplied by the COM LL ratio.

Question 1-5

What are the lengths of the common rafters for R9 considering this 4x ridge beam is set low to allow for lapping rafters above?

RCS pgs. 58-59

Answer: **232.03** (birdsmouth heelcut to birdsmouth heelcut)

Math: 360 -3.5, ÷2, x1.3017 =**232.03**

Method:

Same as done in previous questions except rafter length denotes heelcut to heelcut instead of heelcut to ridgecut.

Question 1-6

What are the common rafter lengths for R1 considering they butt to a 4" thick steel ridge beam?

Answer: **290.28**

Math: 450 -4, ÷2, x1.3017 =**290.28**

Question 1-7

a) *What are TOR heights above the outside wall height for ridges R2, R3, R4, R5 and R7 considering all have common rafters butting to a 4x ridge beam?*

RCS pg. 120

Answers: 1) R2 = **130.80**

2) R3 = **110.80**

3) R4 = **108.30**

4) R5 = **100.80**

5) R7 = **95.80**

Math:

1) 300 -3.5, ÷2, x.8333, +7.26 =**130.80**
2) 252 -3.5, ÷2, x.8333, +7.26 =**110.80**
3) 246 -3.5, ÷2, x.8333, +7.26 =**108.30**
4) 228 -3.5, ÷2, x.8333, +7.26 =**100.80**
5) 216 -3.5, ÷2, x.8333, +7.26 =**95.80**

Method: Span less thk of ridge, divided by 2, multiplied by the RR ratio, plus common rafter heelstand equals TOR.

b) *What is the TOR height above the outside wall height for 4x R9 which is set low to carry lapping rafters above?*

RCS pg. 58

Answer: **148.54**

> *Math:* 360 -3.5, ÷2, x.8333 =**148.54**
>
> *Method*: Same as done in previous question except do not add the common rafter heelstand.

Question 1-8

a) *With rafters butting the ridge beam, R1 is a 4"x20" tubular steel beam welded atop a 4" tubular steel post and fabricated with ears positioned at the ends of each side to receive H1-H4 (5.125 x16 GLM). Draw and calculate the length of this ridge.*

RCS pg. 120

Answer: **23.25** (See **Figure 1-8a**)

> *Math:*
> 1) 462 -450 =12
> * 2) 1.4142 x5.125 =7.25
> 3) 12 +7.25, +4 =**23.25**
>
> *Method*:
> 1) Bldg length less bldg width equals theoretical length.
> 2) Calculate 45° thk hip.
> 3) Theoretical ridge length plus 45° thk hip, plus thk of ridge, equals ridge length.

---

* Book note: 1.4142 is the secant of a 45° right triangle (RCS pg. 301). It is used like the LL ratios to find the 45° diagonal distance when one of the triangle legs are know. For example, in Question 1-8 the 5.125 thk of the GLM is a known leg.

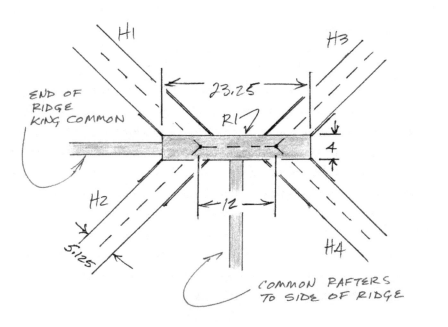

Fig. 1-8a

b) Considering the previous described steel ridge scenario, what is the TOR height above the outside wall height for R1?

Answer: **190.07**

        Math: 462 -23.25, ÷2, x.8333, +7.26, =**190.07**

        Method:

            In this unique steel ridge situation the end-of-ridge king commons would be used to set the ridge height because the hips join the ridge on the sides rather than the ends as is the norm. Therefore, use bldg length instead of span in the ridge height equation shown in Q1-7a.

Question 1-9

a) What is the length of the end-of-ridge king commons for R1?

Answer: **285.56**

        Math: 462 -23.25, ÷2, x1.3017 =**285.56**

b) Considering the previous described steel ridge scenario, how much will the standard common rafters (which butt to the sides of the 4" steel ridge) be above TOR?

Answer: **3.02**

        Math: 3.62 x.8333 =**3.02**

        Method:

            Difference in effective runs between the two varieties of common rafters (223 -219.38 =3.62) multiplied by the RR ratio.

Question 1-10

What is the length of the hips H1-H4?

RCS pgs. 63-67

Answer: **366.05**

        Math: 450 -4, ÷2, x1.6415 =**366.05**

        Method:

            Span less thk of ridge, divided by 2 and multiplied by the H/V LL ratio.

Question 1-11

a) What length would the seatcut be to provide the GLM hip beam with full bearing on the 2x6 wall (consider 6" total wall thickness)? Full bearing in this case is defined as a chevron shaped birdsmouth notch where the plate has 100% contact with the hip along the sides of the seatcut. There will be a small triangle of overcut inside the building.

Answer: **8.49**

        Math: 6 x1.4142 =**8.49**

b) Referring to a full plate width seatcut on the 5.125x16 hip, what would be the associated heelcut dimension at the side of hip where it crosses the outside plate line?

Answer: **5**

> *Math*: 6 x.8333 =**5**
> *Method*: Wall thickness multiplied by the RR ratio.

c) For the 5.125x16 hip to have no more than a full bearing seatcut and yet plane in with the 2x8 common rafters, a beam pocket will need to be made in the outside wall. How deep should it be made for this hip in a non-backed situation?

Answer: Wall lowered = **6.31**

> *Math*:
>
> SOLVED using CM calculator.
> 10 Rise, 16.9706 Run, Pitch = Pitch, 16 Run, Diag <*18.57*> -7.26, -5 =**6.31**
>
> *Method*:
>
> Use the CM calculator to find the pitch of the hip when using roof rise per 12" of roof run (10) as the rise leg together with the hip travel per 12" of roof run (16.9706) as the run leg. Apply this pitch to calculate the hip's plumb dimension. Then from this dimension subtract off the common rafter heelstand and the hip's heelcut dimension. The result is the amount needed to lower the wall.

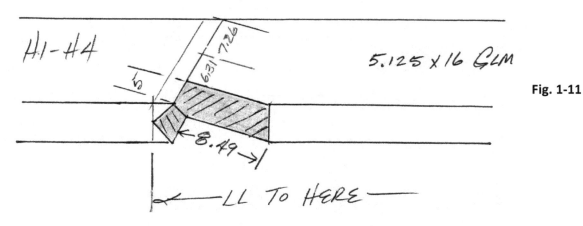

Fig. 1-11

Question 1-12

What is the saw bevel-angle if the GLM hip beams were to be backed? How far down measured perpendicular from the top edge of the hip would one locate the rip-cut line?

RCS pgs. 67-69, 187

Answer: **26.92** degrees; **1.30** rip-cut line

> *Math*:
>
> SOLVED using CM calculator.
> 10 Inch Pitch, 12 Run, Hip/V <*19.70*> = Run, 10 Rise, Pitch Pitch (**26.92**) = Pitch, 5.125 ÷2, = Run, Rise (**1.30**)
>
> *Method*:
>
> Use the CM calculator to calculate the angle equivalent when the hip LL for 12" of roof run (19.70) is set as the run leg and the roof rise for 12" of roof run (10) is set as the rise leg. Use this angle to find the rip-cut line by substituting ½ thk of the hip as the run leg and solve for the rise leg.

Question 1-13

How should one cut the birdsmouths for H3 and H4?

RCS pg. 105

Answer: The H3 and H4 birdsmouths are cut tailless following the double 45° heelcut V-notch.

Question 1-14

How are the lower ends of H3 and H4 located?

Answer: With both H3 and H4 cut tailless following the double 45° heelcut V-notch, position the tip of H3's seatcut flush with the outside of the 270" long wall (R5) at a measurement of 120" from the adjacent gable end corner. This will match the inside corner where SV2 terminates. Double check bldg width (450) at this position back to the inside corner at BV1. For H4, flush the tip of H4's seatcut in line with the outside wall running from the H2 corner to BV1 at a measurement of 462" from the H2 corner (See **Figure 1-14**).

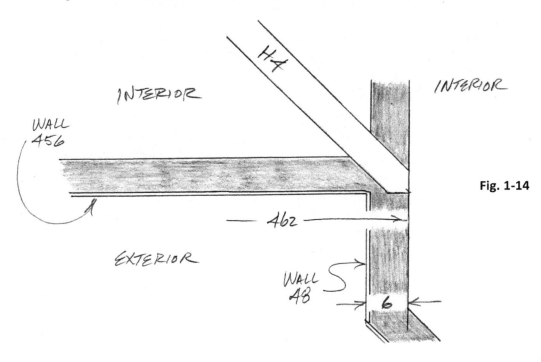

Fig. 1-14

Question 1-15

What are the lengths of the supported valleys SV1 – SV7?

RCS pgs. 100-107

Answers: 1) SV1 = **243.25**

2) SV2 = **184.16**

3) SV3 = **174.31**

4) SV6 = **282.65**

5) SV4 = **175.25**

6) SV5 = **208.80**

7) SV7 = **199.88**

Math:

Prelim calcs

   A) 1.4142 x5.125, ÷2 =3.62

   B) 1.4142 x3.5, ÷2 =2.47

Regular calcs

   1) 300 -3.62, ÷2, x1.6415 =**243.25**

   2) 228 -3.62, ÷2, x1.6415 =**184.16**

   3) 216 -3.62, ÷2, x1.6415 =**174.31**

   4) 246 +102, -3.62, ÷2, x1.6415 =**282.65**

   5) 216 -2.47, ÷2, x1.6415 =**175.25**

   6) 252 -2.47, ÷2, x1.6415 =**208.80**

   7) 246 -2.47, ÷2, x1.6415 =**199.88**

Method:

   A) ½ 45° thk of 5.125 GLM

   B) ½ 45° thk of 4x

   1-7) Span less ½ 45° thk supporting member divided by 2 and multiplied by the H/V LL ratio.

Note:

   For supported valleys hanging from 5.125 GLM hips (SV1, SV2, SV3, SV6) use ½ 45" thk of 5.125 in the equation. For supported valleys (SV4, SV5, SV7) hanging from 4x material use ½ 45" thk of 3.5 in the equation.

Question 1-16

How are the upper ends of the supported valleys SV1 – SV7 cut?

Answer: Square-cut plumb, at the H/V pitch.

Question 1-17

How are the heelcuts made on SV3 and SV5? How would one locate where to nail the birdsmouth when stacking for each of these cases?

RCS pg. 102

Answer: 1) For SV3, a single cheek 45° heelcut is made (as opposed to a standard V-notch birdsmouth) centered on the lower LL plumb-line. When stacked, center SV3 on a measurement 216" from the H2 outside corner.

          2) SV5 is cut tailless with a single 45° heelcut centered on the lower LL plumb-line. It would stack to the interior continuation of wall 74 (don't forget to compensate for the lack of ½" sheer as opposed to the outside portion of the wall) centered on a measurement of 156" (74 + 82) from the outside gable end corner of R3.

Question 1-18

What is the backing angle for the supported valleys if this process is desired?

RCS pg. 67-69, 90

Answer: Same as calculated previously for the 5.125" GLM hip (26.92 degrees). No cut-line is required on a valley since the backing angle slopes inwards (concave) from the top outside edges.

Question 1-19

When stacking, how would one find the correct position where the SV rafters would hang from their supporting members?

RCS pgs. 105-107

Answer: Side-to-side alignment

Position the SV rafters to their associated supporting rafter by measuring up the supported rafter from the lower LL plumb-line (at the birdsmouth) the full theoretical H/V length for the SV's corresponding span. A plumb-line at this measurement would mark the center of where the SV rafter attaches. The supporting rafters should be laid out with these position marks prior to being stacked.

1) SV1 to H4 = **246.23**
2) SV2 to H3 = **187.13**
3) SV3 to H2 and SV4 to H6 = **177.28**
4) SV5 to H6 = **206.83**
5) SV6 to H2 = **285.62**
6) SV7 to SV6 = **201.90**

Math:

1) 300 ÷2, x1.6415 =**246.23**
2) 228 ÷2, x1.6415 =**187.13**
3) 216 ÷2, x1.6415 =**177.28**
4) 252 ÷2, x1.6415 =**206.83**
5) 246 +102, ÷2, x1.6415 =**285.62**
6) 246 ÷2, x1.6415 =**201.90**

Up-and-down alignment

For a SV to a supporting hip, align the center of the SV's top edge at the headcut (middle of the 3.5" beam width dimension in these cases) with the edge of the hip from which it hangs. For a SV to a supporting valley (SV7 to SV6 and SV4 to H6), align the top inside edge (edge closest to R4, R6) with the top edge of the rafter from which it hangs.

Question 1-20

a) What is the lengths of H5/H6 if they are cut to butt head-to-head against the end of 4x R2?

Answer: **246.23** to the tip of a double 45° cheekcut.

Math: 300 ÷2, x1.6415 =**246.23**

Method: Use full span for calculations. Do not subtract out the thk of ridge.

b) What is the length of H7/H8 if they are cut to butt head-to-head and sit on top of 4x R9?

RCS pgs. 75-79

Answer: **295.47** to the tip of a double 45° cheekcut.

Math: 360 ÷2, x1.6415 =**295.47**

c) How do H5/H6 and H7/H8 differ at the top cut?

Answer: H7/H8 have a birdsmouth for the 4x ridge beam on which they sit. H5/H6 do not.

Question 1-21

a) *Rafter H6 is an unusual rafter in that it serves as both a hip and valley. How is this rafter laid out?*
   RCS pg. 107

Answer: H6 is laid out as a valley rafter with respect to birdsmouth design and modified into a hip near the top. If it is not to be backed, the birdsmouth would be laid out as a normal valley positioning the common rafter heelstand in the center of the rafter at the lower LL plumb-line. That portion of the rafter between R6 and R3 must be backed for a hip on the side adjacent SV4. That portion of the rafter from R3 to R2 must be backed for a hip on both sides. Backing is required because of its transition from valley to hip. Recall that in an unbacked hip the birdsmouth is positioned so the two top edges line up with the plane of the roof, whereas in an unbacked valley the center of the top edge lines up with the plane of the roof. Therefore, as the valley changes into a hip that centerline now signifies the surface of the roof plane and the two top corner edges will be above the roof plane requiring them to be cut (backed).

b) *What is the backing angle and rip-cut line location for those parts of H6 where backing is required?*
   RCS pg. 67-69, 104, 107

Answer: **26.92°** degrees; **.89** rip-cut line

   Math:
   SOLVED using CM calculator.
   10 Inch Pitch, 12 Run, Hip/V =Run, 10 =Rise, Pitch Pitch (**26.92**) = Pitch, 3.5 ÷2, = Run, Rise (**.89**)
   Method: Solved as shown earlier except 3.5 (width of 4x) is substituted for the 5.125 shown.
   Note:
   If H6 is backed for a valley then the upper hip section would use the center of the concave valley backing slot as the peak of the convex hip backing. This would put the rip-cut line along the side of the rafter for the hip section at twice the normal depth (1.78). See **Figure 1-21b**.

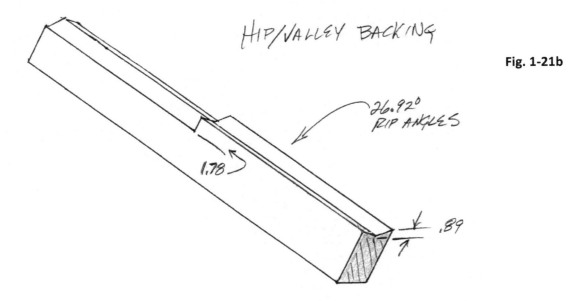

HIP/VALLEY BACKING

Fig. 1-21b

26.92°
RIP ANGLES

1.78

.89

Question 1-22

a) *What are the lengths of R3, R5 and R7 exclusive of an overhang dimension and before making any required adjustments to recenter the ridge (in cases of different material thickness between supporting and supported members)?*

RCS pg. 124

Answers: 1) R3 = **279.53**

           2) R5 = **230.38**

           3) R7 = **104.38**

           Math:

              1) 252 ÷2, +156, -2.47 =**279.53**

              2) 228 ÷2, +120, -3.62 =**230.38**

              3) 216 ÷2, -3.62 =**104.38**

           Method:

              3) ½ span plus bldg extension dimension less ½ 45° thk of the supporting hip.

b) *What type of cut is made on the interior ends of R2, R3, R5 and R7 to allow them to fit in the joint between the supporting hip and SV?*

RCS pgs. 104-105

Answer: A double 45° cheekcut. This cut must be adjusted if there is a difference in thickness between the supported and supporting members (R2, R5, R7).

Question 1-23

Explain the modification one must make to the interior ends of R2, R5 and R7 to correctly center the ridge in their spans?

RCS pg. 104-105

Answer: Add half the difference between the thicknesses of the two rafters (5.125 -3.5, ÷2 =.81) to the side of the ridge beam's double 45° cheekcut that will rest against the 4x SV.   See **Figure 1-23**.

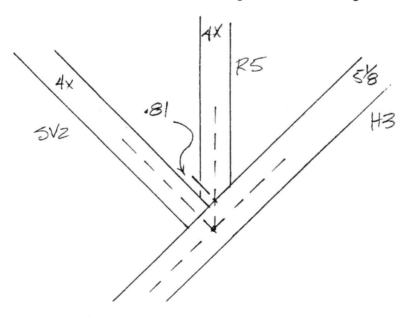

Fig. 1-23

Question 1-24

During stacking, what is the correct height to position the interior ends of R2, R3, R5, R7?

RCS pg. 106-107

Answer: The correct ridge height for these offshooting ridges is when the top corner edge abutting the hip planes in with the top far edge of the hip from which it hangs.

Question 1-25

a) What is the length of R2 (considering H5 and H6 butt head-to-head) prior to adjusting for the difference in material thickness between H4 and SV1?

RCS pgs. 124-127

Answer: **324.38**

Math: 328 -3.62 =**324.38**

Method: Length of associated wall less ½ 45° thk H4.

b) How is the end of R2 that butts to H5/H6 cut?

Answer: Square-cut.

Question 1-26

a) Referring to the 4/12 shed dormer originating near R9, what is the height of the dormer's front wall (above the existing plate height) considering the roof run is 144" and the rafters are 2x8s with a 3" seatcut ?

RCS pgs. 192-193

Answer: Wall height = **72.35**

Math:
1) 7.5 x1.0541 =7.91
2) 3 x.3333 =1
3) 7.91 -1 =6.91
4) 10 -4 =6
5) 144 x.5000 =72
6) 72 -6.91, +7.26 =**72.35**

Method:
1) Calculate 4/12 plumb dimension of 2x8.
2) Calculate heelcut dimension.
3) Subtract heelcut dimension from 2x8 rafter plumb dimension to calc heelstand.
4) Calculate dormer rate of closure (6/12).
5) Calculate ht at 144" from point of roof convergence using rate of closure pitch.
6) Adjust for difference in heelstands (4/12 vs 10/12) to find wall ht.

b) *What are the lenghts of the shed dormer's common rafters and the 10/12 pitch upper cripple jacks if the dormer's headout is a dbl 2x fitted into the 10/12 pitch roof plane?*

Answers: Shed dormer commons = **151.79**

10/12 pitch cripple jacks = **40.68**

Math:

1) 144 x1.0541 =**151.79**
2) 360 -3.5, ÷2, -144, -3, x1.3017 =**40.68**

Method:

1) Run multiplied by the 4/12 COM LL ratio.
2) Span less thk ridge, divided by 2, less shed dormer run, less thk headout, multiplied by the 10/12 COM LL ratio.

Question 1-27

a) *What is the length of 2x WV1?*

Answer: **256.48** shortened 1.5" (to fit in the 45° corner at intersection of R9 and wall 329).

See **Figure 1-27a**.

Math: 360 -3.5, ÷2, -22, x1.6415 =**256.48**

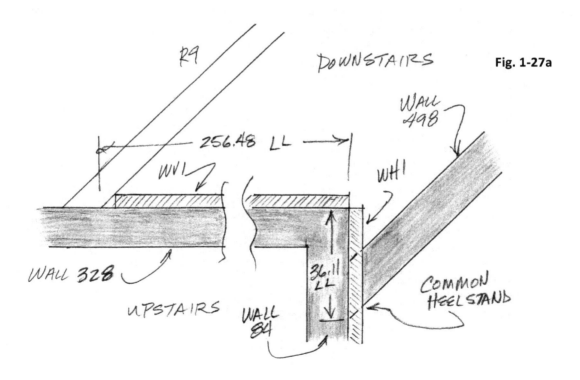

Fig. 1-27a

b) How is WV1 cut and stacked?

Answer: Since WV1 is a nailed-in-place wall ledger, I would not waste time with a birdsmouth at R9 (remember R9 is a 4x set low) but cut it to fit between R9 and the wall corner by H5. Cut both ends square and plumb at the H/V pitch. Position WV1 by snapping a chaulk-line denoting top of common rafters for R9 on the wall and nail the ledger below this line. Jack rafters will plane in with the top edge of WV1, that is against the wall (the snapline, in other words). The wall snapline can be located at the top by measuring up plumb at the near edge of R9 the common rafter heelstand dimension. At the bottom, locate the snapline position on the bldg wall corner below H5 by measuring up from the lower wall plates the rise for the 22" run plus the heelstand dimension (22 x.8333, +7.26 =25.59).

c) What is the length of WH1, which is an around-the-corner continuation of ledger WV1?

RCS pgs. 238-239 (similar)

Answer: **36.11**

Math: 22 x1.6415 =**36.11**

Method: Ledger WH1 is a hip so the 22 run is multiplied by the H/V LL ratio.

Note: WV1 and WH1 together should equal the length of H8 if it were cut to butt R9.

d) How is WH1 cut and stacked?

Answer: The rafter is measured from the headcut plumb-line to the LP of a single 45° heelcut at wall 498 (tail runs wild). The headcut is square-cut plumb at the H/V pitch. The heelstand of a common rafter is positioned on the side face of WH1 that is away from wall 84 at the point where it crosses the wall 498 (SP of the single cheek 45° heelcut). When stacking, the lower end of WH1 is automatically positioned correctly against the wall on account of the birdsmouth. At the top, position the edge away from the wall to plane in with an imaginary continuation of the line snapped for WV1 along wall 328.

Question 1-28

On lower roof R9, how much higher is wall 72 than the main plate height of wall 461.75?

RCS pg. 11, 14-15

Answer: **56.79**

Math: 68.15 x.8333 =**56.79**

Method: Wall run multiplied by the RR ratio.

Question 1-29

a) How can wall 150 best be constructed?

RCS pgs. 248-251

Answer: Construct wall 150 as a rake wall substituting a backed 4x6 beam for the two top plates. This method allows the jack rafters to have plate bearing without making a complicated birdsmouth.

b) What is the height at each end of rake wall 150 (SP and LP above R9's plate height)?

Answer: SP wall = **56.79**; LP wall = **145.17**

        Math: 150 ÷1.4142, x.8333, +56.79 =**145.18**

        Method:

            SP Because rake wall 150 begins from the corner intersection with wall 72, this ht is used for the SP.

            LP Wall 150 runs at 45° to the roof pitch (diagonally) or the same as does a H/V. Convert the 150 wall
                length into a run dimension and solve for rise. Use that rise together with the SP height to calculate LP.

c) What is the length of the top plate beam for rake wall 150?

Answer: **174.11** measured along outside top edge

        Math: 150 ÷1.4142, x1.6415 =**174.11**

        Method: Convert the 45° diagonal wall 150 length into a run dimension and solve for a H/V length.

d) How would one layout and cut backing on the top plate beam?

    RCS pg. 250

Answer: A backing cut is made from outside the wall to modify the finished rake wall for a no birdsmouth condition.
        This backing cut is the reason that it is best to use a solid top plate beam rather than double top plates. Mark a
        backing rip-cut line by measuring down plumb at each outside end of the rake wall the common rafter heelcut
        dimension and snap a line. Make a rip-cut following this line using the saw set to the angle found earlier for
        backing H/Vs (26.92 degrees).

e) What layout along the top of wall 150 will place the rafters at 24" OC?

    RCS pg. 75 (similar)

Answer: **39.40**

        Math: 24 x1.6415 =**39.40**

        Method: OC spacing multiplied by the H/V LL ratio.

f) How does one layout and cut plumb frieze blocks for the rafters bearing on wall 150?

Answer: Block length = **36.93**. Ends are cut with paralleling 45° cheekcuts following H/V plumb-lines.
        See **Figure 1-29**.
        Math: 22.50 x1.6415 =**36.93**
        Method: Spacing between rafters multiplied by the H/V LL ratio.

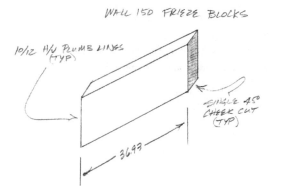

Fig. 1-29

Question 1-30

a) What is the best method to frame the bay roof associated with 4x R4?

RCS pgs 231-237

Answer: The bay roof associated with R4 is an equal-lateral style or half an octagon fitted on the end of a straight gable roof extension. Start by butting 2 theoretical length common rafters head-to-head at the end of the 4x ridge and have a single common rafter running off these at 90° to the middle of the center bay wall OH2-OH3. Two additional common rafters would run off the middle of each of the other two bay walls (wall OH3-OH4, wall OH1-OH2) to the top connection. Octagon hips OH1-OH4 would be cut to fit in the gaps between these 5 common rafters and frame to the corners. See **Figure 1-30a**.

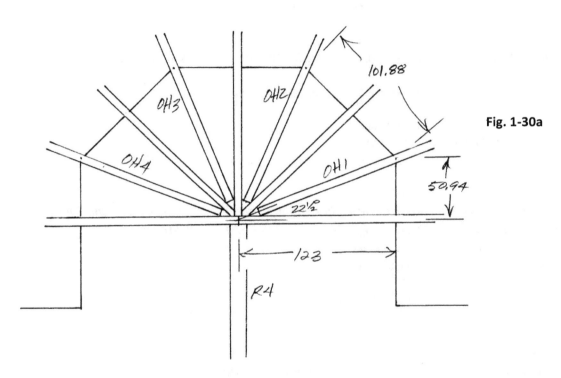

Fig. 1-30a

b) Considering the previous described method, what are the lengths of the various rafters for the bay roof at R4?

RCS pg. 232

Answers: 1) Head-to-head commons at end of R4 = **160.11**

2) Common to center of wall OH2-OH3 = **159.13**

3) Commons to walls OH3-OH4 and OH1-OH2 (to tip of dbl 45° cheekcut at top) = **158.73**

4) Octagon hips **9.25**/12 pitch (to a square-cut top) = **163.20**

Math:

1) 246 ÷2, x1.3017 =**160.11**

2) 246 -1.5, ÷2, x1.3017 =**159.13**

3) 246 -2.12, ÷2 x1.3017 =**158.73**

4) SOLVED using CM calculator.

10 Inch Pitch, 246 ÷2 Run, Rise <*102.5*> Stor 1, 22.5 Pitch, Diag <*133.13*> = Run, Rcl 1 = Rise, Pitch Pitch Pitch Pitch (**9.25**) =Pitch, Run -3.82 =Run, Diag (**163.20**)

Method:

1) Run multiplied by the COM LL ratio.

2) Run less ½ thk butting commons multiplied by the COM LL ratio.

3) Run less ½ 45° thk butting commons multiplied by the COM LL ratio.

4) Solve a rt triangle of ½ bldg jog angle (22.5°) for the hypotenuse (hip travel) when bldg run is set as the run leg. Use this hip travel (133.13) together with the theoretical roof rise (102.5) to calculate the hip theoretical length and shorten for a square-cut fit at the head.

Note:

The 3.82" measurement used to shorten the hip run for a square-cut top is found by solving for the radius of a sixteen 2x rafter circumference (16 x1.5, ÷6.28 =3.82). RCS pg. 223

c) *Where should one layout the common rafter heelstand at the OHs' birdsmouths?*

RCS pg. 235

Answer: **.31"** uphill measured level from the LL plumb-line at the birdsmouth.

Math:

SOLVED using CM calculator.

22.5 Pitch, .75 Run, Rise (**.31**)

Method: Solve a rt triangle of ½ bldg jog angle (22.5°) for the rise leg when ½ thk OH is set as the run leg.

d) *Where along the wall 120 extensions, in relation to the bldg corners at OH1 and OH4, would one position the two head-to-head king commons framed to the end of the R4?*

RCS pg. 159 (similar)

Answer: Centered on **50.94** from the OH1 and OH4 corners.

Math: 101.88 ÷2 =**50.94**

Method: ½ the bay roof wall dimension.

Question 1-31

a) *What is the 24" OC jack step length for the bay roof at R4?*

RCS pgs. 151-154

Answer: **75.42**

Math:

SOLVED using CM calculator.

22.5 Pitch, 24 Rise, Run = Run, 10 Inch Pitch, Diag (**75.42**)

Method:

Solve a rt triangle of ½ bldg jog angle (22.5°) for the run leg when the OC spacing is set as the rise leg. Convert this dimension into a LL measurement.

b) *What angle would one set a swing-table saw to make these jack rafter headcuts if one opted to do so, rather than cutting them square and plumb at the SP?*

RCS pg. 151, 156 - 157

Answer: **67.5°**

Math: 90 -22.5 =**67.5**

Method: 90° less ½ bldg jog angle.

Question 1-32

What is the length of R4 considering the previously described bay roof top connection which has two commons butting head-to-head and one common running off these at 90°?

Answer: **188.84**

      Math: 246 ÷2, +120, -50.94, -.75, -2.47 =**188.84**

      Method:

          Span divided in half plus bldg extension, less side wall position of butting king commons, less ½ thk 2x common, less ½ 45° thk 4x SV6.

Question 1-33

a) How would one frame the bay roof associated with 2x R6?

      RCS pgs. 231-237

Answer: Similar as was done for the bay roof at R4 except the two opposing king commons will frame to butt the ridge and the roof pitch for the two front 45° corner wall sections (BP1-BP2, BP3-BP4) will not be 10/12.

b) What are the various rafter lengths for this roof if framed similarly to what was done for the bay roof at R4?

      RCS pg.233

Answers: 1) Three king commons, two to walls 72 and one to center of wall BP2-BP3 = **139.61**

          2) BP1-BP4 hips (tops are square-cut plumb) = **148.66** pitch **8.70**/12

          3) Common rafter to center of walls 66 (tops to LP of dbl 45° cheekcut) = **148.47** pitch **9**/12

      See **Figure 1-33b.**

      Math:

          1) 216 -1.5, ÷2, x1.3017 =**139.61**

          2&3) BP1-BP4 and the common rafters centered on walls 66 are SOLVED using the CM calculator in progressive steps (similar to RCS pg 233) for clarity rather than fully utilizing the calculator's memory capacity to shorten the calculations. Clear the calculator before each step.

          A) 216 ÷2 Run, 61.33 Rise, Diag (124.20)

          B) 10 Inch Pitch, 216 ÷2 Run, Rise (90)

          C) 124.20 Run, 90 Rise, Pitch (**8.70**) =Pitch, Run -3.82, =Run, Diag (**148.66**)

          D) 124.20 Diag, 33 Rise, Run (119.74)

          E) 119.74 Run, 90 Rise, Pitch (**9**) =Pitch, Run -1.06, =Run, Diag (**148.47**)

      Method:

          A) Calculate hip theoretical travel.

          B) Calculate rise at theoretical top framing point on R6.

          C) Use rise and travel to calc BP1-BP4 lengths (shortened for the 3.82 as shown in Q1-30b) and pitch/12.

          D) Calculate common rafter run to wall 66.

          E) Use rise and run to calc common rafter length to wall 66 (shortened for ½ 45° thk 2x) and pitch/12.

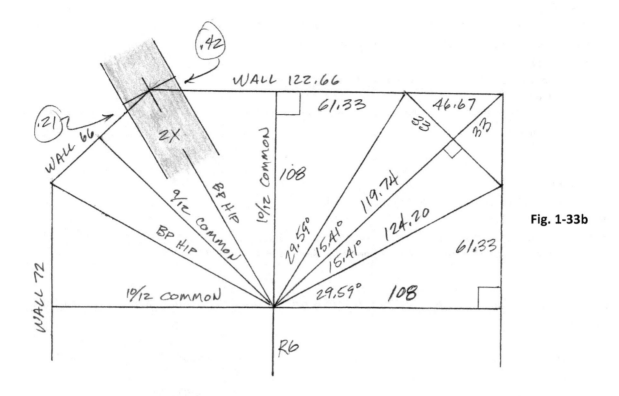

Fig. 1-33b

c) *Where is the common rafter heelstand marked on the 2x BPs? What is the difference in heelstands between the two sides?*

RCS pg. 235-237

Answer: 1) The common rafter heelstand is marked on the 10/12 side of the BP rafter measured (**.43**) horizontally uphill from the LL plumb-line at the birdsmouth. The matching horizontal measurement on the 9/12 side of the rafter is (**.21**).

2) The difference in heelstand heights between sides is (**.16**)

Math:

SOLVED using CM calculator.

1) 108 Run, 61.33 Rise, Pitch Pitch <*29.59°*> = Pitch, .75 Run, Rise (**.43**), 45 - Pitch, = <*15.40°*> Pitch, Rise (**.21**)

2) **Off**, On/C, 8.7 Inch Pitch, .43-.21, Run, Rise (**.16**)

Method:

The heelstand of a common rafter is marked on the 10/12 side of the BP at that point where it crosses the exterior wall line. One should verify that the heelstand on the off-pitch side of the rafter is within .125". If not, the hip should be backed or slid off center. This height difference can be calculated mathematically by using the horizontal difference between the two sides' heelstands (.43-.21 =.22) as a run measurement in combination with the pitch of the the BP to find the height difference from side to side. One can also make a top view scale drawing of the hip corner, then layout a piece of lumber with the hip's birdsmouth to compare the difference in heelstands from side to side. Usually, with 2x hip material the difference is close to the .125" max tolerance - as is true in this case (.16).

d) *Where along the wall 72 extensions, in relation to the bldg corners at BP1 and BP4, would one position the two opposing king commons framed at the end of R6?*

      RCS pg. 159 (similar)

Answer: Centered on **61.33** from the BP1 and BP4 corners.

      Math: 122.66 ÷2 =**61.33**

      Method: ½ the bay roof end wall (122.66) dimension.

Question 1-34

a) *What are the two different jack step lengths needed for the bay roof at R6?*

      RCS pgs. 167-170 (similar)

Answers: 1) **55.01** for 10/12 pitch side at wall 122.66 and walls 72

      2) **108.94** for 9/12 pitch side at walls 66

      Math:

          SOLVED using CM calculator.

          1) 61.33 ÷24 =<*2.56*> Stor 1, 108 Run, 10 Inch Pitch, Diag <*140.58*> ÷Rcl 1 =**55.01**

          2) 33 ÷24 =<*1.38*> Stor 1, 119.74 Run, 90 Rise, Diag <*149.79*> ÷Rcl 1 =**108.94**

      Method:

          Divide half the end and corner Bay roof wall lenghts by the OC spacing to find the number of jacks needed. Then divide the associated theoretical common rafter lengths by these numbers to calculate the step lengths.

b) *What angle would one set a swing table saw to make these jack rafter headcuts if one opted to do so, rather than cutting them square plumb at the SP?*

Answers: 1) **60.41**° for jacks from wall 122.66 and walls 72

      2) **74.59**° for jacks from walls 66

      Math:

          Prelim calcs

          29.59 and *15.41* SOLVED using CM calculator.

          A) 108 Run, 61.33 Rise, Pitch Pitch <*29.59*°> =Pitch, 45 –Pitch, =15.41°

          Regular calcs

          1) 90 -29.59 =**60.41**

          2) 90 -15.41 =**74.59**

      Method:

          Solve the rt triangle for it's angle when using the bldg run set as the run leg and ½ Bay roof end wall length set as the rise leg. Subtract the result (29.59°) from the bldg jog angle (45°) to find the opposing angle (15.41°). Subtract each individually from 90° to get the saw's bevel-cut angles.

Question 1-35

a) *What is the length of R6 considering the previously described bay roof top connection which had two opposing king commons butting the end of the 2x ridge and one common running off the end of the ridge at 90°?*

Answer: **116.95**

        Math: 216 ÷2, +72, -61.33, +.75, -2.47 =**116.95**

        Method:

                Shown previously in Q1-32, except .75 is added rather than subtracted because the two opposing king commons at the end of the ridge run to walls 72, butt the side of the ridge. This extends the ridge ½ thk of the 2x rafter past the theoretical framing point (similar to RCS pg. 70).

b) *What is the correct up/down position of the interior ends of R4 and R6 where they hang from their supporting members?*

        RCS pg. 105

Answer: Correct up/down position is where the two top corner edges of the ridge plane into the center of the adjacent valley rafter on each side. This holds true for the left side of R6 (side which butts H6) as well, for at this point it is playing the role of a valley rafter.

Question 1-36

a) *How would one frame the bay roof associated with R8?*

        RCS pgs. 231-237

Answer: This is another equal-lateral bay roof similar to the one at R4 except there are no wall returns for the principal span. BV1 and BV2 will be CA framed over the lower end of SV5 and adjacent common rafters. The common rafters from R3 continue inside the bldg under the bay roof and stack tailless on a continuation of wall 74 to SV5. The common rafters from R2 also continue inside the bldg under the bay roof to SV5 and are cut short to hang from a ceiling beam that spans from the end of H4 to end of SV5. One end of 2x R8 is cut to rest directly on non-backed SV5, while the other end butts to a common rafter running off the center of wall OH6-OH7. Common rafters are run off the center of walls OH5-OH6 and OH7-OH8. Their tops are cut with a dbl 45° cheekcut to nail to the end-of-ridge common. The 4 OH hip tops are square-cut plumb as was done for the bay roof at R4. OH5 and OH8 are cut tailless to stop at the BV1 and BV2 valley corners. False valley tails are installed off BV1 and BV2 to handle the eave transition (RCS pgs. 134-135). Off-angle bastard CA valley fill is run from R8 to snaplines at BV1 and BV2.

See **Figure 1-36a**.

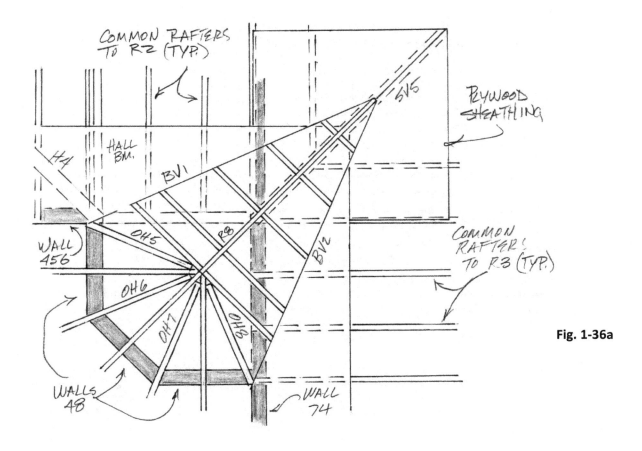

Fig. 1-36a

b) *What are the various rafter lengths and ridge height for this bay roof?*

Answers: 1)   Common rafter to wall OH6-OH7 = **74.50**

2)   Common rafters to walls OH5-OH6 and OH7-OH8 (tops to LP of dbl 45° cheekcut) = **74.10**

3)   OH5-OH8 hips (tops are square-cut plumb) = **74.35**, pitch **9.25/12**

4)   Ridge height = **54.95**

Math:

Prelim calcs

A) 82 x1.4142, =115.97

calculate bay roof span

Regular calcs

1) 115.97 -1.5, ÷2, x1.3017 =**74.50**

2) 115.97 -2.12, ÷2, x1.3017 =**74.10**

3) SOLVED using CM calculator.

22.5 Pitch, 48 ÷2 Rise, Diag Stor 1 <*62.72*>, 115.97 ÷2 Run, 10 Inch Pitch, Rise <*48.32*> = Rise, Rcl 1 Run, Pitch (**9.25**) =Pitch, Run -3.82, =Run, Diag (**74.35**)

4) 115.97 -1.5, ÷2, x.8333, +7.26 =**54.95**

Method: See Q1-30b.

Question 1-37

What is the length of R8 if framed to bear on non-backed SV5?

Answer: **114.89**

> *Math*:
>
> SOLVED using CM calculator in steps for clarity. Clear the calculator before each step.
> 1) 45 Pitch, 48 Diag, Rise (33.94)
> 2) 115.97 -1.5, ÷2 =57.24
> 3) 22.5 Pitch, 57.24 Rise, Run <*138.19*> +33.94, -57.24 =**114.89**
>
> *Method*:
> 1) Use wall 48's dimension as the diagonal of 45° rt triangle to calculate the bay roof's projection (33.94) from a span line drawn from the OH5/BV1 inside corner to the OH8/BV2 inside corner
> 2) Use the bay roof's span less thickness of ridge, divided by 2 to calculate effective run (57.24).
> 3) Use the effective run as the rise leg of a 22.5° rt triangle and solve for the run leg. To this dimension (138.19) add the bay roof's projection then subtract off the effective run of the common rafter originating from the OH6-OH7 wall and butting to the end of R8. See **Figure 1-37**.
>
> *Note*:
>
> Because this Bay roof dormer is positioned bisecting a 90° inside bldg corner, BV1 and BV2 are at ½ 45° (22.5°) in relation to R8.

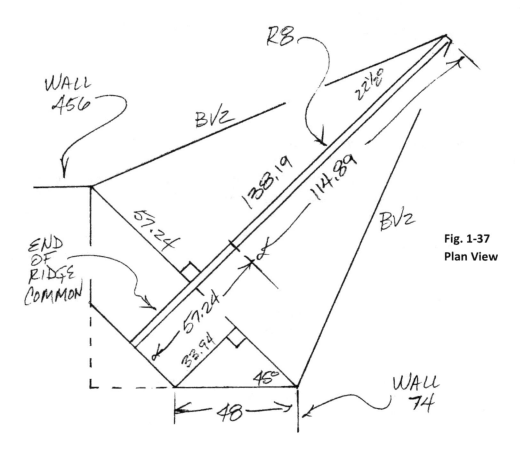

Fig. 1-37
Plan View

Question 1-38

a) What is the jack step length for the CA valley jack rafters spanning from R8 to BV1/BV2?

Answer: **12.94**

> *Math:*
>
>> SOLVED using the CM calculator.
>>
>> 22.5 Pitch, 24 Run, Rise =Run, 10 Inch Pitch, Diag (**12.94**)
>
> *Method:*
>
>> Solve a 22.5° rt triangle for the rise leg when OC spacing is set as the run leg. Convert this dimension into a LL measurement.

b) What is the sidecut and bottom edge angle-cut for these CA valley jacks?

> RCS pgs. 80, 242-245

Answers: 1) **70.31°** sidecut (19.69° off of square across board) or **4.25**/12 for Quick Square® usage.

2) bottom edge angle-cut = **26.92°**

> *Math:*
>
>> SOLVED using CM calculator.
>>
>> 1) 10 Inch Pitch Pitch, Stor 1, 10 Rise, 16.9706 Run, Pitch Pitch <*30.51°*> + Rcl 1 =(**70.31°**), -90, = conv +/- <*19.69°*> , Pitch Pitch Pitch Pitch Conv 8 (**4.25**)
>>
>> 2) 10/12 H/V backing angle
>
> *Method:*
>
>> 1) Add the level-cut angle for a 10/12 pitch common rafter together with the level-cut angle for a 10/12 pitch hip (these rafters sit at 45° to the roof plane surface).
>>
>> 2) The bottom edge angle-cut is equal to the backing angle on a 10/12 H/V (26.92°) since the jack rafters run 45° to 10/12 roof plane. Calculating the H/V backing angle was shown in Q1-12.

c) What is the length of the longest CA valley jacks (to LP at lower end) that serve as the side king commons at the bay roof end of R8 considering ½ " roof plywood and a 2x sleeper are installed?

Answer: **59.56**

> *Math:*
>
>> SOLVED using CM calculator in steps for clarity rather than utilizing the calculator's memory
>>
>> 1) 10 Inch Pitch, 2 Rise, Diag <*3.12*> =Run, 45 Pitch Diag (4.42)
>>
>> 2) 114.89 -4.42, = Run, 22.5 Pitch, Rise = Run, 10 Inch Pitch, Diag (**59.56**)
>
> *Method:*
>
>> 1) Calculate the hypothetical amount the ridge would be shortened for the 2" thicknesses of roof plywood and sleeper. Solve using the 10/12 pitch and convert to a hip travel dimension.
>>
>> 2) Subtract this hypothetical amount from the ridge length and set the result as the run leg in a 22.5° rt triangle (what R8 forms with BV1/BV2) then solve for the rise leg (end-of-ridge common rafter effective run). Convert this dimension into a LL measurement. See **Figure 1-38c**.

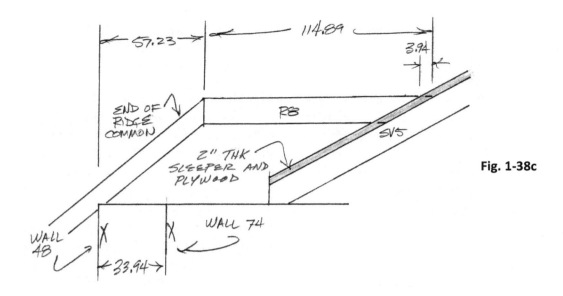

Fig. 1-38c

Question 1-39

What is the length of the end-of-wall commons SC1 and SC2 at R7?

Answers: 1) SC1 = **133.59** to LP 45° cheekcut.

2) SC2 = **135.09** to LP 45° cheekcut.

Math:

1) 216 -3.5, ÷2, -3.62, x1.3017 =**133.59**

2) 216 -3.5, ÷2, -2.47, x1.3017 =**135.09**

Method: common rafter LL less ½ 45° thk hip/valley.

Note:

SC2 will require small seatcut (3 -2.47 =.53") at the bottom corner edge of the LP 45° cheekcut.

See **Figure 1-39**.

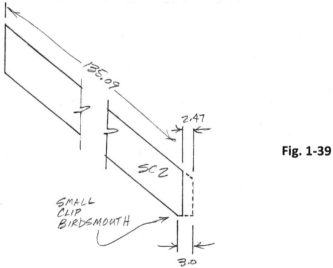

Fig. 1-39

Question 1-40

What is the regular 24" OC hip/valley jack step length for Roof 1?

RCS pg. 69

Answer: **31.24**

       Math: 24 x1.3017 =**31.24**

       Method: OC spacing multiplied by the COM LL ratio.

Question 1-41

a) How many sets of regular hip/valley jacks are needed for this roof?

Answer: Hip fill

        left set of 3 -lower end H5

        full set of 4 -lower end H1/H2

        full set of 6 -H3/H5

        2 full sets of 7 -H7,H8

        2 left sets of 9 -H1/H4

    Valley fill

        full set of 4 -R7

        2 full sets of 5 -R5,R6

        2 full sets of 6 -R4,R3

        full set of 7 -R2

        full set of 7 w/birds top -R9

        left set of 7 -R2

> **Book note:** A "full jack set" includes both left and right side jacks.

b) RCS demonstrates a method of gang cutting a full set of jack fill by pairing the left and right side jacks together. What can be done if one ended up needing various ½ sets as in the above cut list for this roof?

RCS pgs. 71, 92

Answer: Pair the unmatched ½ sets of the same side together (left hip w/left hip, etc) or intermix an unpaired of one type with an unpaired of other (left set valley w/ unpaired left set hip). Just remember to coordinate the direction of the cheekcuts in the center correctly. See **Figure 1-41b**.

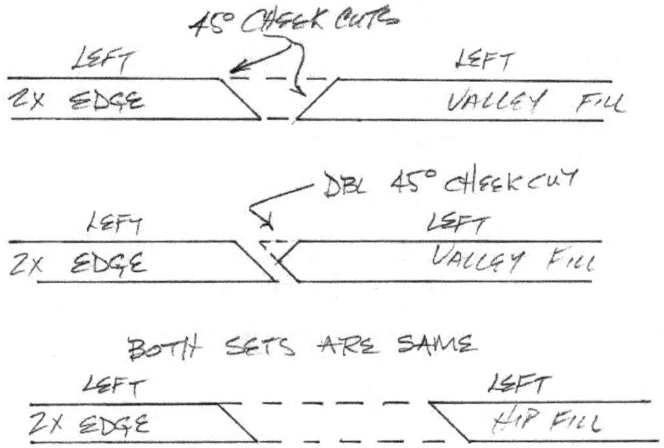

Fig. 1-41b

Top view of jacks as gang cut on racks. Compare with RCS Figures 4-8, 5-6

Question 1-42

a) *What is the shortening amount for the various sized hips one would incorporate into the hip jack calculations to step down from a king common?*

RCS pg. 68-70

Answer: 1) **4.72** for 5.125 GLM hips.

2) **3.22** for 4x hips.

Math:

1) 5.125 ÷2, x1.4142, x1.3017 =**4.72**

2) 3.5 ÷2, x1.4142, x1.3017 =**3.22**

b) *How is this shortening amount applied when marking layout for hip jacks stepping down from a common rafter?*

RCS pg. 71-72

Answer: Beginning with the corresponding common rafter length subtract off the appropriate shortening dimension for the hip size (calculated above) together with one jack step length to mark out the longest hip jack to SP. From that longest hip jack length progressively shorten each successive rafter by one jack step length. This procedure is done during rafter layout and requires no calculation.

Question 1-43

a) *If the hip jacks from H3 to the outside wall are cut to start from a hip corner (similar to H1), what is the length of the longest continuation jack (ie: valley jacks stacked to a hip) to LP?*

Answer: **107.40**

Math:

1) 5.125 x1.4142, x1.3017 =9.43

2) 24 -1.5, x1.3017 =29.29

3) 146.12 -29.29, -9.43 =**107.40**

Method:

1) LL of 45° thk 5.125 GLM.

2) 1st hip jack LL converted to SP.

3) Common rafter length for R5 less 1st hip jack length to SP, less LL of 45° thk hip.

b) *How do you position the lower end of these continuation jacks from R5 to H3?*

RCS pg. 105

Answer: Held high to plane in with the far top edge of the hip across from where they attach.

Question 1-44

What are the lengths of the various parallel hip-valley jacks in Roof 1? Approximately how many of each size are required?

RCS pgs. 73-74

Answer: 1) H1-SV6/ H2-SV7 (GLM/4x) = **124.85**   10 jacks

2) H3-SV1 (GLM/4x) = **187.33**   4 jacks

3) H5-SV4 (4x/4x) = **102.91**   4 jacks

Math:

    Prelim calcs

        A) 5.125 ÷2, x1.4142 =3.62

        B) 3.5 ÷2, x1.4142 =2.47

    Regular calcs

        1) 102 -3.62, -2.47, x1.3017 =**124.85**

        2) 270 -120, -3.62, -2.47, x1.3017 =**187.33**

        3) 84 -2.47, -2.47, x1.3017 =**102.91**

Method:

    A) Calculate ½ 45° thk 5.125 GLM.

    B) Calculate ½ 45° thk 4x.

    1-3) Hip to valley run less ½ 45° thk hip and valley, multiplied by the COM LL ratio.

Question 1-45

What is the 24"OC layout up/down the hips and valleys to position parallel hip-valley jacks when stacking?

RCS pg. 75

Answer: **39.40**

    Math: 24 x1.6415 =**39.40**

    Method: OC spacing multiplied by the H/V LL ratio.

Question 1-46

a) What length is the 24" OC diverging hip-valley jack step for Roof 1?

RCS pg. 73-74

Answer: **62.48**

    Math: 24 x1.3017, x2 =**62.48**

    Method: Twice the regular 24"OC H/V jack step length.

b) How should one cut and stack sets of diverging hip-valley jacks for speed and efficiency?

    RCS pg. 74

Answer: 1) Cut the shortest jack as one step size in length from LP to LP and progress larger by one step size for each succeeding jack. Remember the LP of the 45° cheekcut at each end are on the same side of rafter. The side with the LPs depending on whether is a left or right fill.

           2) To stack, mark the 24" OC hip/valley layout (39.40 calculated in Q1-45) upward on a hip or downward on a valley beginning from the inside intersection of the hip with the valley.

c) How many sets of diverging hip-valley jacks are required in Roof 1?

Answer: left set of 2 -H2/SV6

           rt set of 1 -H6/SV4

           rt set of 2 -SV6/SV7

           rt set of 3 -H4/SV1

           2 rt set of 4 -H2/SV3, H3/SV2

Now, with all these questions answered one has the information needed to cut the roof for Plan 1. But before we move on to Roof 2, I added one "what if" hypothetical question to illustrate a common roof variation.

Question 1-47

If the plate height for the bay roof extension at R4 was raised 12" higher than the adjacent main wall plate height, what would change?

RCS pgs. 98-99

Answer: If we consider the eave overhang would stay equal to the main house, one would now have and over/under roof situation. The changes would be as follows:

1) The lower eaves would terminate against walls OH1-SV6 and OH-SV7 while the upper eaves would die on top of the H1-H2 roof plane surface.
2) Both SV6 and SV7 would move outward towards their adjacent hip corners, be longer in length and cut tailless at the H1-H2 wall line.
3) The parallel hip-valley jacks would be shorter.
4) One less diverging jack would be needed from H2 to SV6 and SV6 to SV7.
5) One less hip jack rafter would be needed at the lower ends of H1 and H2 on the side of the roof towards bay roof extension.
6) A small continuation jack rafter from both SV6 and SV7 to the H1-H2 wall line would be required to carry the short overframed wall extensions (14.40") of walls OH1-SV6 and OH2-SV7. A plywood nailer would need to be attached to these jacks so the roof sheathing for the H1-H2 roof plane can continue over to the new inboard wall extensions.
7) The lower ends of SV6/SV7 below the OH1-SV6 and OH2-SV7 walls must be backed on inside top edge to fit into the new continuation of H1-H2 roof plane. See **Figure 1-47** and RCS pg. 107.
8) Ridge would be longer by the run equivalent for 12" of roof rise.

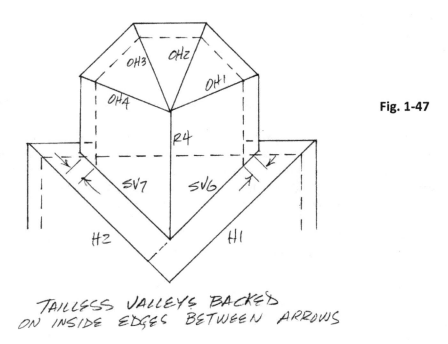

Fig. 1-47

b) *What are the new rafter lengths required for this variation?*

Answers: 1) SV6 = **294.47**

2) SV7 = **223.51**

3) SV6-H1/SV7-H2 parallel hip-valley jacks = **106.10**

4) Small hip jack to SP at wall OH1-SV6 and wall OH2-SV7 = **15.53**

Math:

Prelim calcs

A) 12 ÷.8333, =14.4

Regular calcs

1) 246 +102, +14.4, -3.62, ÷2, x1.6415 =**294.47**

2) 246 +28.8, -2.47, ÷2, x1.6415 =**223.51**

3) 102 -14.4, -3.62, -2.47, x1.3017 =**106.10**

4) 14.4 -2.47, x1.3017 =**15.53**

Method:

A) Calculate run for 12" rise.

1) R4 span plus distance to H2 corner, plus run for 12" rise, less ½ 45° thk H2, divided by 2 and multiplied by the H/V LL ratio.

2) R4 span plus twice the calculated run for the 12" rise (14.40), less ½ 45° thk SV6, divided by 2, and multiplied by the H/V LL ratio.

3) Wall distance from the H1 and H2 corners to center of the SVs (102 -14.40), less ½ 45° thk of both the SV (2.47) and GLM hips (3.62), with the result multiplied by the COM LL ratio.

4) Distance to center of SV's from the inside corner (run for 12" rise or 14.40") less ½ 45° thk SV, with the result multiplied by the COM LL ratio.

Roof 2 – Phase 1
Monte Sereno CA
6/12 pitch

All ridges, hips, valleys are 2x except as noted

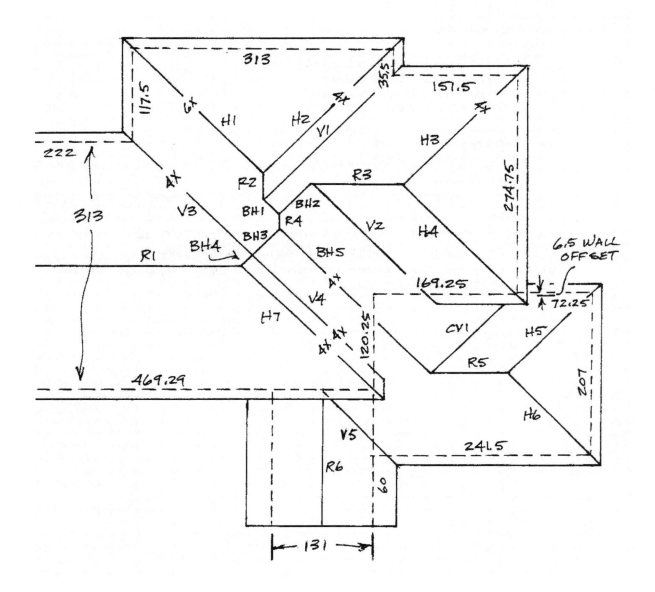

ROOF 2

I divided Roof plan 2 vertically down the middle and solved each half separately. The right side encompasses a chopped up regular hip roof with some interesting features. This is **Phase 1**. The left side encompasses a unique angular hip roof overwhelming with challenges in both visualization and calculation. This is **Phase 2**.

Phase 1

A man's reach should exceed his grasp.

Question 2-1

What is the RR ratio for this 6/12 pitch roof?

RCS pg. 3, 299

Answer: **.5000**

   Math: 6 ÷12 = **.5000**

Question 2-2

What are the common rafter and hip/valley LL ratios for this 6/12 roof?

RCS pgs. 3, 4, 298

Answers: COM LL ratio = **1.1180**

            H/V LL ratio = **1.5000**

Question 2-3

Considering the roof pitch is 6/12, the rafters are 2x8 and the seatcut is 3", what is the heelstand?

RCS pgs. 1, 53

Answer: **6.89**

        Math:

            1) 3 x .5000 = 1.5
            2) 7.5 x 1.1180, - 1.5 =**6.89**

Question 2-4

What is the length of the common rafters for R1, R2, R3, R5 and R6 considering they all butt to 2x ridge boards?

RCS pgs. 49-50

Answers: 1)   R1, R2 = **174.13**

2)   R3 = **152.75**

3)   R5 = **114.88** regular commons

4)   R5 = **105.93** tailless common rafters run under CV1 to a 2x ledger against wall 169.25

5)   R6 = **72.39**

Math:

1)   313 -1.5, ÷2, x1.1180 =**174.13**

2)   274.75 -1.5, ÷2, x1.1180 =**152.75**

3)   207 -1.5, ÷2, x1.1180 =**114.88**

4)   207 -1.5, ÷2, -6.5, -1.5, x1.1180 =**105.93**

5)   131 -1.5, ÷2, x1.1180 =**72.39**

Method:

4) From the effective run subtract the thickness of the dbl sided sheathed wall 169.25 (6.5) and a 2x ledger (1.5).

Question 2-5

a) What are TOR heights above the outside wall height for ridges R1, R2, R3, R5, R6 considering they all have common rafters butting to 2x ridge boards?

RCS pg. 120

Answers: 1)   R1, R2 = **84.77**

2)   R3 = **75.20**

3)   R5 = **58.27**

4)   R6 = **39.27**

Math:

1)   313 -1.5, ÷2, x.5000, +6.89 =**84.77**

2)   274.75 -1.5, ÷2, x.5000, +6.89 =**75.20**

3)   207 -1.5, ÷2, x.5000, +6.89 =**58.27**

4)   131 -1.5, ÷2, x.5000, +6.89 =**39.27**

b) What is the height of the 2x ledger along wall 169.25 which carries the lower end of the tailless R5 common rafters that are installed for a cathedral ceiling below CV1?

Answer: **10.14** above lower wall height to top of ledger. Plane rafters in with the wall side of the ledger.

Math: 6.5 x.5000, +6.89 =**10.14**

Method:

The distance from the outside face of wall 72.25 to the outside face of wall 169.25 (6.5), multiplied by the RR ratio, the result added to the heelstand dimension of a common rafter (6.88).

Question 2-6

What is the TOR dimension for R4? Explain how this dimension is found.

Answer: **99.76**

Math: 373 -1.5, ÷2, x.5000, +6.89 =**99.77**

Method:

The span for R4 stretches from an imaginary interior "continue thru" line from wall 117.5 to where V2 crosses wall 169.25.

To find this span one must start by solving for the difference in plate heights between the downstairs walls for R5 downstairs and the upstair walls. Notice V3 and V4 butt head-to-head. This signifies that the span from the wall where V4's birdsmouth is positioned on wall 120.25, measured left over to an imaginary interior continuation of wall 117.5 (where V3's birdsmouth is located) is equal to the span from wall 222 down to an imaginary horizontal line drawn drawn parallel to wall 469.29 from where V4 sits on the outside wall. In other words, the 45° line denoting V3-V4 is the diagonal of a imaginary square box defined by these bearing points. See **Figure 2-6-1**.

By finding the difference between the lengths of wall 169.25 and wall 151.5 and subtracting this dimension (17.75) from the wall 313 dimension, the span for V3-V4 is found (295.25). Knowing that the roof plane from V5-H6 is in line with the bottom end of V4 we have enough info to solve the lower wall height.

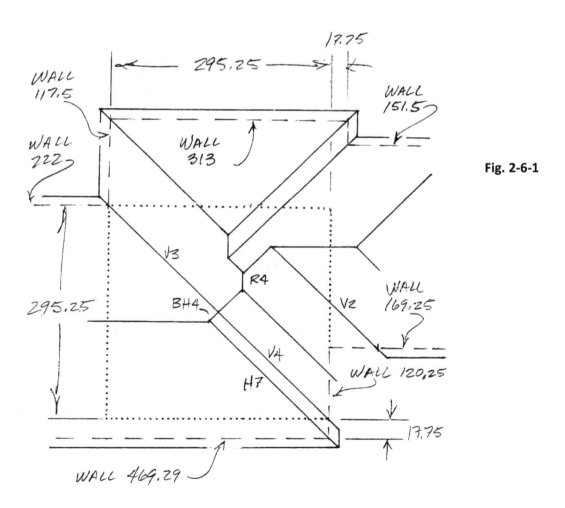

Fig. 2-6-1

Taking into account that span 207 (downstairs) starts 6.5" past wall 169.25 we can do the math to solve for the distance from wall 241.5 to wall 469.29 (207 -6.5, -120.25 =80.25). If the distance from wall 469.29 (H7 corner) along wall 120.5 to the center of V4 is 17.75" to match the 295.25 span we calculated earlier, the run distance from wall 241.5 to center birdsmouth of V4 is 98" (80.25 +17.75 =98). Calculating the rise for this run and we have the plate height difference between the upstairs and downstairs walls (98 x.5000 =49).

Working with this plate height difference we are now able to calculate where V2 sits on wall 169.25. CV1 begins at the bldg corner H4 (low). Therefore, while it is not readily visible, the rise for the roof plane R4-V2 begins at the outside edge of wall 72.25 (see **Figure 2-6-2**).

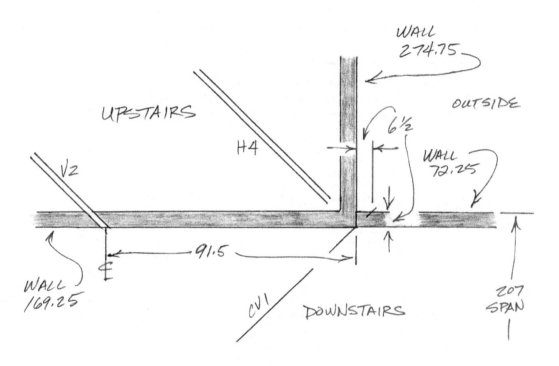

Fig. 2-6-2

Technically speaking wall 72.25 is the plate line for CV1. Imagine if CV1 was run as a regular valley, it must reach wall 72.25 as does the 98" run measurement corresponding to the 49" difference in plate heights. After subtracting the 6.5" of run from outside surface of wall 72.25 to the H4 corner (wall sheathed both sides with ½" ply), the remaining 91.5" dimension (98 -6.5) is measured from the H4 corner along wall 169.25 to the center of V2 (see **Figure 2-6-3**). By subtracting this distance from the length of wall 169.25 (169.25 -91.5 =77.75) we have the increase in span distance over the V3-V4 span (295.25 +77.75 =373) for R4.

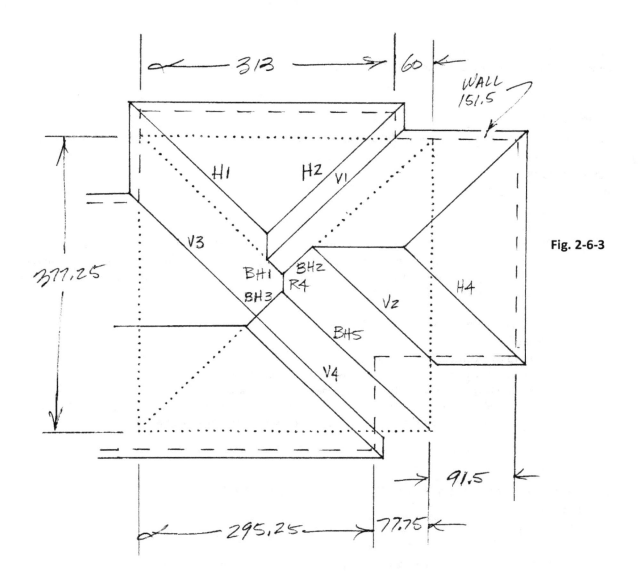

Fig. 2-6-3

Question 2-7

a) What are the lengths of hips/valleys H1, H2, V1, H7, H3, H4, V2, H5 and H6 considering they all butt to 2x ridge boards? What are the lenghts of V3 and V4 considering they butt head-to-head?

RCS pgs. 63-66, 87-88

Answers: 1) H1, H2, V1, H7 = **233.63**

2) H3, H4, V2 = **204.94**

3) H5, H6 = **154.13**

4) V3, V4 = **221.44**

Math:

1) 313 -1.5, ÷2, x1.5000 =**233.63**
2) 274.75 -1.5, ÷2, x1.5000 =**204.94**
3) 207 -1.5, ÷2, x1.5000 =**154.13**
4) 295.25 ÷2, x1.5000 =**221.44**

b) How are the lower ends of V4 and V2 cut?

Answer: Tailless with single or double 45° cheekcut following the heelcut V-notch.

Question 2-8

What are the theoretical ridge lengths of R2, R3 and R4? Would any of the ridges be cut to length? Why or why not?

RCS pg. 63, 73-75

Answers: 1) R2 = **35.5**

2) R3 = **91.5**

3) R4 = **4.25**

4) None of these ridges would be cut to length. In all cases it would be best to run them wild but laid out with the above theoretical lengths as an aid to stacking. R2 and R3 should be run long on the interior ends for their associated broken hip connections, while R4 should be run long at both ends to avoid the need to cut special king hip-valley jacks.

Math:

3) 377.25 -373 =**4.25**

Method:

1) Length to match parallel wall 35.5

2) Length to match distance from H4 corner to V2 (found earlier).

3) Bldg length less bldg width. The bldg length for R4 is measured from an imaginary interior continuation of wall 151.5 to an imaginary interior horizontal line thru the birdsmouth of V4 (313 +117.5, -35.5, -17.75 =377.25). The bldg width is the span dimension found earlier in Q2-6 (373).

Question 2-9

What is the length of R5 if it runs past BH5 and 6" into wall 120.25?

RCS pgs. 63-64

Answer: **144.75**

Math: 241.5, -(207÷2), +.75, +6 =**144.75**

Method: Total bldg extension less ½ span, plus ½ thk ridge, plus wall penetration.

Question 2-10

What are the lengths of BH1, BH2, BH5 considering 2x ridges?

RCS pgs. 78-79

Answers: 1) BH1 = **42.75** (See **Figure 2-6-3**)

2) BH2 = **71.44** (See **Figure 2-6-3**)

3) BH5 = **269.25** (See **Figure 2-10**)

Math:

1) 373 -313, ÷2, -1.5, x1.5000 =**42.75**

2) 373 -274.75, ÷2, -1.5 x1.5000 =**71.44**

3) 373 +(2 x98), -207, ÷2, -1.5, x1.5000 =**269.25**

Method:

Major span less minor span, less ½ thk top and bottom ridges and multiplied by the H/V LL ratio.

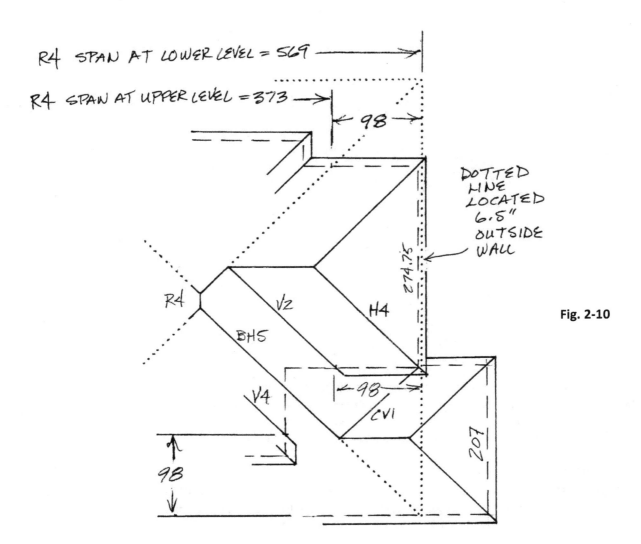

R4 SPAN AT LOWER LEVEL = 569

R4 SPAN AT UPPER LEVEL = 373

98

DOTTED
LINE
LOCATED
6.5"
OUTSIDE
WALL

274.75

R4 √2 H4

BH5

V4

98

CV1

98

207

Fig. 2-10

Question 2-11

a) What are the lengths of BH3 and BH4 considering they butt to the 4x V3/V4 connection?

Answer: 1) BH3 = **55.33**

2) BH4 = **10.33**

Math:

Prelim calcs

A) 3.5 ÷1.4142, ÷2 =1.24

Conversion of V3/V4 4x thk to a hypothetical ridge thk to fit the formula (see **Figure 2-11a**)

Regular calcs

1) 373 -295.25, ÷2, -.75, -1.24, x1.5000 =**55.33**

2) 313 -295.25, ÷2, -.75, -1.24, x1.5000 =**10.33**

37

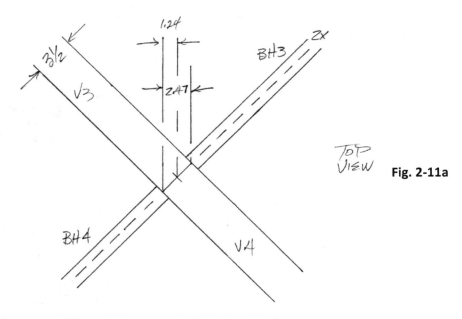

Fig. 2-11a

b) How are BH3 and BH4 cut differently from standard broken hips?

Answer: Standard broken hips have a dbl 45° cheekcut top and bottom. BH3, BH4 have a dbl 45° cheekcut top and are square-cut plumb at the bottom (to nail to V3, V4).

c) How are the lower ends of BH3 and BH4 stacked?

Answer: Center the lower ends of BH3 and BH4 on the head-to-head joint between V3/V4 so the top outside edges of the broken hips die into the middle of the 4x valleys on each side (see **Figure 2-11c**).

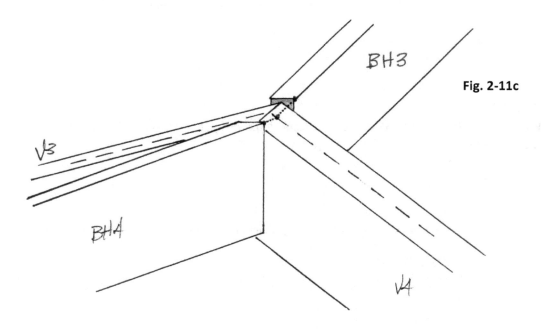

Fig. 2-11c

Question 2-12

a) What is plate height difference between wall 241.5 and wall 60?

RCS pg. 100-102

Answer: **7**

> *Math*: 80.25 -1.5, -(131 -1.5, ÷2), x.5000 =**7**
>
> *Method*:
>
>> Notice that V5 tucks into the inside corner formed by wall 469.29 and R6. This shows that R6 planes with the V5-H6 roof plane at this point. With a 2x wall ledger installed on wall 469.29 (to catch valley jacks from V5) the distance from wall 241.5 to this ledger is 78.75" (80.25 [from Q2-6] -1.5). The difference between this measurement and the effective run of R6 (131 -1.5, ÷2 =64.75) results in a run measurement of 14". The rise for that run is 7" (14 x.5000 =7).

b) What is the length of V5?

Answer: **118.13**

> *Math*: 80.25 -1.5, x1.5000 =**118.13**
>
> *Note*:
>
>> A second birdsmouth would be positioned uphill 21" LL (14 x1.5000 =21) to bear on wall 60. Both birdsmouth's heelcuts are made with a single 45° cheekcut (RCS pg. 102 similar). See **Figure 2-12b**.

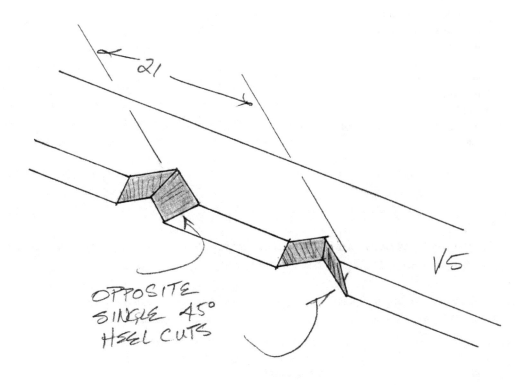

Fig. 2-12b

c) *How far out along wall 241.5 from the inside corner with wall 60 should one position the lower birdsmouth of V5?*

RCS pg. 102

Answer: **14**

Method: Difference in runs, calculated in Q2-12a.

Question 2-13

What is the regular 24" OC hip/valley jack step length for Roof 2?

RCS pg. 70

Answer: **26.83**

Math: 24 x1.1180 =**26.83**

Question 2-14

How many sets of regular hip/valley jacks are needed for the Phase 1 part of the roof?

RCS pg. 73

Answer: Hip fill

    left set of 1 -lower end H2

    2 left sets of 3 -lower end H4, BH5 to R5 (CA cut to bear on 2x sleeper at R5. See Q2-21d)

    rt set of 4 -lower end H1

    2 full sets of 4 -H5,H6

    full set of 5 -H3

    rt set of 5 -H4

    left set of 6 -H7

    full set of 6 -H1/H2

  Valley fill

    rt set of 3 -V2 (start from valley/ridge as opposed to stepping down from common)

    full set of 4 -V5

    left set of 6 -V1

    rt set of 7 -V3

Question 2-15

What are the lengths for various parallel hip-valley jacks in Phase 1 of this roof? Approximately how many of each size are required?

RCS pgs. 73-74

Answers:

1)	H2-V1 [4x-2x] = **35.74**	7 jacks (1 field cut at R2)
2)	H1-V3 [6x-4x] = **124.25**	8 jacks (2 field cut at R2)
3)	H4-V2 [2x-2x] = **99.93**	2 jacks
4)	H7-V4 [4x-4x] = **14.32**	7 jacks
5)	BH5-V2 [4x-2x] = **110.65** (see **Figure 2-15-1**)	4 jacks (1 field cut at R4)
6)	BH5-V4 [4x-4x] = **81.40** (see **Figure 2-6-3**)	5 jacks
7)	BH1-V3 [2x-4x] = **87.73** (see **Figure 2-15-2**)	3 jacks (1 field cut at R4)
8)	BH2-V1 [2x/2x] = **64.71** (see **Figure 2-15-2**)	2 jacks

Math:

Prelim calcs

A) ½ 45° 6x =3.89

B) ½ 45° 4x =2.47

C) ½ 45° 2x =1.06

Regular calcs

1) 35.5 -2.47, -1.06, x1.1180 =**35.74**

2) 117.5 -3.89, -2.47, x1.1180 =**124.25**

3) 91.5 -1.06, -1.06, x1.1180 =**99.93**

4) 17.75 -2.47, -2.47, x1.1180 =**14.32**

5) 377.25 -274.75 =102.5 (calculate run distance)

102.5 -2.47, -1.02, x1.1180 =**110.65**

6) 373 -295.25 =77.75 (calculate run distance)

77.75 -2.47, -2.47, x1.1180 =**81.40**

7) 117.5 -35.5 =82 (calculate run distance)

82 -1.06, -2.47, x1.1180 =**87.73**

8) 373 -313 =60 (calculate run distance)

60 -1.06, -1.06, x1.1180 =**64.71**

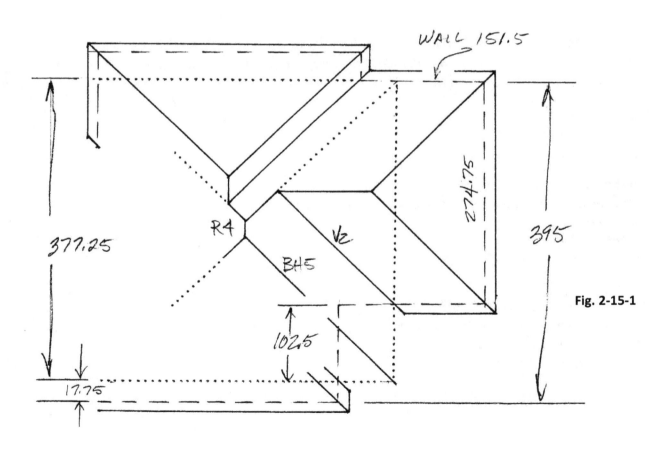

Fig. 2-15-1

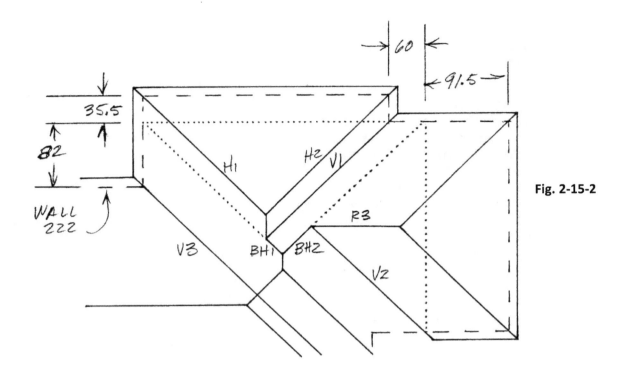

Fig. 2-15-2

Question 2-16

What is the 24"OC layout up/down the hips and valleys to position parallel hip-valley jacks when stacking?

RCS pg. 75

Answer: 36

> *Math*: 24 x1.5000 =**36**

Question 2-17

What length is the regular 24"OC diverging hip-valley jack step for Roof 2?

RCS pg. 73

Answer: 53.66

> *Math*: 24 x2, x1.1180 =**53.66**

Question 2-18

How many sets of diverging hip-valley jacks are required in Phase 1 of this roof?

Answer: 2 left set of 1 -BH1/V1, BH3/V3
left set of 2 -BH2/V2
rt set of 1 -BH3/V4
rt set of 4 -BH5/CV1 (CA cut lower ends to bear on a 2x valley sleeper at CV1)

Question 2-19

What are some reasons NOT to cut end-of-ridge king common parallel hip-valley jacks?
RCS pg. 73

Answer: In chopped up roofs it is easiest to run the ridges long and slap a regular parallel hip-valley jack to the side of the ridge. Avoiding "one time only" special rafters substantially lessens workload and negates the likelihood of loosing them during stacking.

Question 2-20

a) In this roof, where might one cut diverging king hip-valley jacks to fill in open spaces (ie: not currently filled with a parallel or diverging hip-valley jack)?
RCS pg. 73-76

Answer: From R4-V3, from R4-V2 and from R2-V3 (See **Figure 2-23**).

b) In reality, rather than cutting these "one time only" diverging king hip-valley jack rafters, what would one do to keep things simple for stacking and save time?

Answer: Cut a couple extra parallel hip-valley jacks for H1-V3, BH1-V3, BH5-V2 and have the stackers use these to field cut the 3-4 jacks needed to fill in the open areas (See Q2-15 and **Figure 2-23**) .

c) FOR PRACTICE ONLY, calculate the diverging king hip-valley jacks for R4 (at BH2)-V2 and R4 (at BH3)-V3.
RCS pg. 76

Answers: 1) R4 (at BH2)-V2 = **107.82** (square-cut head at the ridge to center of a 45° cheekcut at the lower end)

2) R4 (at BH3)-V3 = **83.32** (square-cut head at the ridge to center of a 45° cheekcut at the lower end)

Math:

1) 373 -274.75, -.75, -1.06, x1.1180 =**107.82**
2) 373 -295.25, -.75, -2.47, x1.1180 =**83.32**

Method:

Subtract off ½ thk of the ridge and ½ 45° thk of the valley from the difference between the major and minor spans. Multiply the result by the COM LL ratio. (See **Figure 2-23**)

Note:

To calculate a jack off the opposite end of R4 for each of the cases given above, multiply the theoretical length of the ridge (ex: R4 =4.25) by the COM LL ratio and add to the measurements found above.

Question 2-21

a) What is the length of the longest diverging hip-valley jack from BH5-CV1 that is nailed against wall 169.75? Consider the lower end will be CA framed (layover intersection) on top of ½" plywood sheathing and a 2x sleeper.

RCS pg. 97-98

Answer: **209.13** from LP of 45° cheekcut at the head to LP of 26½° bevel-cut at the CA cut lower end.

Math:

Prelim calcs

    A) 2 x1.1180, ÷.5000 =4.47

Regular calcs

    1) 207 ÷2, -6.5, x2, -2.47, -4.47, x1.1180 =**209.13**

Method:

A) Convert the combined thickness of the roof sheathing and the sleeper into a plumb dimension (2 x1.1180 =2.24), then find the run corresponding to this rise (2.24 ÷.5000 =4.47).

1) Take the span of the lower roof and divide it in half, then subtract off the 6.5" inward projection of wall 169.25. Double the result (to get theoretical jack run) and subtract off both the ½ 45° thk of BH5 (2.47) and the run dimension found in the prelim calcs corresponding to the 2" thick material set on top of the R5 commons (4.47). Convert this result into a LL measurement by multiplying by the COM LL ratio.

b) During stacking, how should one position the 2x sleeper at CV1?

Answer: Position the upper end of the 2x sleeper so the outside top edge planes in with the adjacent top edge of BH5 at its connection with R5. Locate the lower end of the 2x sleeper by setting it inbound from the wall 169.75/wall 274.75 corner by 4.47" (run dimension associated with the 2" thick material).

See **Figure 2-21b**.

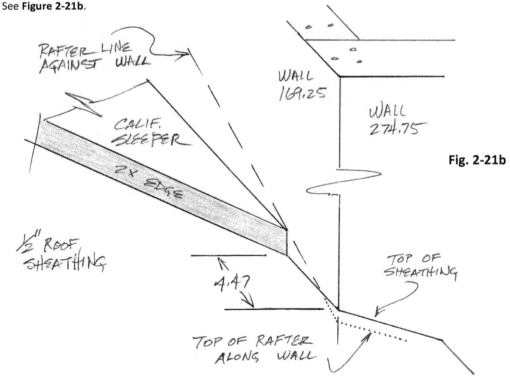

Fig. 2-21b

c) *While it is common practice for the stackers to hand cut the valley sleepers during assembly, what length would CV1 be if precut? How are the upper and lower ends cut?*

RCS pg. 80 (similar)

Answer: **142.15** measured along the top outside edge. The cut at the bottom end against the wall is laid out using a framing square set at 12/13.375 and the saw's bevel-angle set at 26.5° as described in RCS pg. 95. For the cut at the top, position the framing square similarly but mark both sides of the square so you end up with a 12/13.375 rt triangle drawn on the upper end of the sleeper with the 90° corner positioned at the LL mark. Now, connect the mid-point of the hypotenuse on the far side of sleeper back to the 90° corner as shown in the associated drawing. The saw's bevel-angle for this cut would be the same as used for hip backing (18.5°). See **Figure 2-21c-1**.

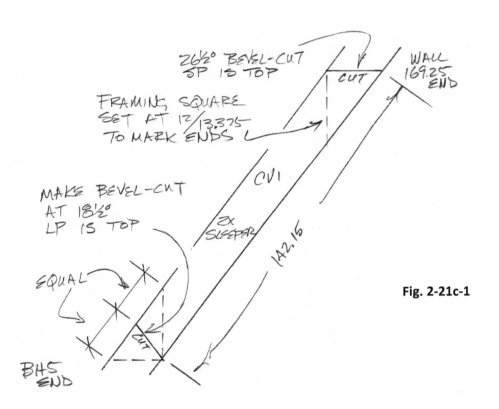

Fig. 2-21c-1

Math:

Prelim calcs

    1.50 SOLVED using CM calculator.

    A) 6 Inch Pitch Pitch, x2, Pitch <*53.13°*>, 2 Rise, Run (1.50)

    B) 2 ÷.5000 =1

Regular calcs

    1) 207 ÷2, -6.5, x1.1180 <*108.45*>, -1.5, -1 <*105.95*>, ÷1.1180 <*94.76*>, x1.5000 = **142.15**

Method:

    A) Using a close up cross-section view of the roof from R5 thru wall 169.25, find the amount the top surface of the sleeper is shorter than the bottom at the upper end by solving a 37° rt triangle for the rise leg when the 2" thk of sleeper and plywood is set as the run leg (See **Figure 2-21c-2**).

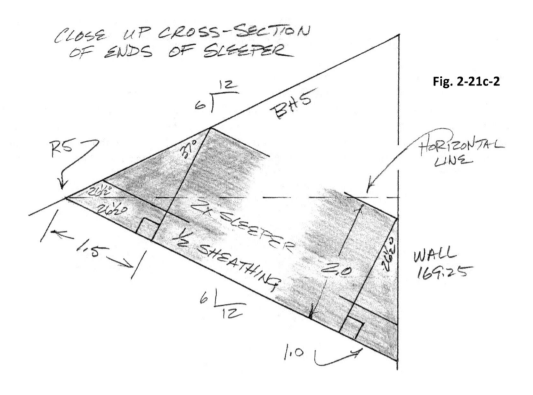

CLOSE UP CROSS-SECTION OF ENDS OF SLEEPER

Fig. 2-21c-2

$6 \overline{|^{12}}$

BH5

R5

37°

HORIZONTAL LINE

26½"

26½0

1.5

2× SLEEPER

½ SHEATHING

2.0

26½0

WALL 169.25

$6 \underline{|_{12}}$

1.0

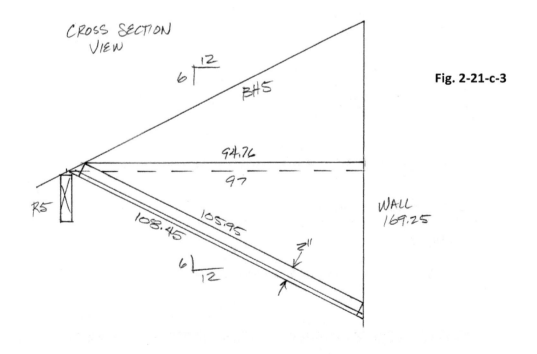

CROSS SECTION VIEW

Fig. 2-21-c-3

$6 \overline{|^{12}}$

BH5

94.76

97

105.95

108.45

R5

3"

WALL 169.25

$6 \underline{|_{12}}$

46

B) Solve for the similar case at the lower end of the sleeper by dividing the 2" material thickness by the RR ratio.

1) Using a cross-section view of the roof from R5 thru wall 169.25, calculate the LL for the top of the 2x sleeper from BH to the wall (or take measurements from a snapped out full-size drawing on the floor). Next, convert this measurement (105.95) to a run dimension by dividing by the COM LL ratio, and finally multiply this run by the H/V LL ratio to find the top of the CA sleeper length (142.15). See **Figure 2-21c-3**.

d) What is the sidecut and bevel-cut angles at the lower ends of the three CA framed hip jacks from BH5 to R5?

RCS pgs. 242 - 244

Answer: **53.13°** sidecut (36.87° off square across the board) or **9**/12 for Quick Square® usage. There is no bevel-cut angle at the lower ends because these are situated at 90° to the roof surface (saw set at **0** degrees).

Math: SOLVED using CM calculator.

6 inch Pitch Pitch, x2 =**53.13°**, -90, =Conv +/- <*36.87°*>, Pitch Pitch Pitch Pitch (**9**)

Method: Double the roof pitch. See **Figure 2-21c-2**.

Question 2-22

a) Why did I choose to cut the lower end of BH5 in the regular fashion (dbl 45° cheekcut) to nail to the side of R5 rather than cut to bear directly on the CV1 sleeper as is shown in RCS pg. 80-81?

Answer: If BH5 was run to bear on a 2x sleeper at CV1, all the commons to R5 would need to be installed and the sheathing run before the sleeper could be set. When stacking, one likes to get all the skeleton (roof outlining rafters) up without having to backtrack and run intermediate commons or plywood. By making BH5 a regular ridge-to-ridge broken hip, it can be stacked in sequence without delay. Besides that, a ridge-to-ridge broken hip connection is easier to calculate, cut and stack than a layover broken hip connection.

b) During stacking, where would one connect the lower end of this regular cut BH5 to R5?

Answer: With a king common positioned off the end of ridge R5, BH5 would be connected **67.25**" from end of R5 (see
Figure 2-22b).

Math:
1) 72.25 -6.5 =65.76
2) 65.75 +.75, +.75 =**67.25**

Method:
1) Find the theoretical length of R5 (65.75) by subtracting the 6.5" valley offset dimension from the wall
72.25 length.
2) Convert the theoretical ridge length to an actual length (67.25) by adding ½ thk ridge (.75) for the end-
of-ridge king common deduction and a second ½ thk ridge (.75) to account for the 45° offset from
theoretical length resulting when a broken hip is nailed to the side of a ridge.

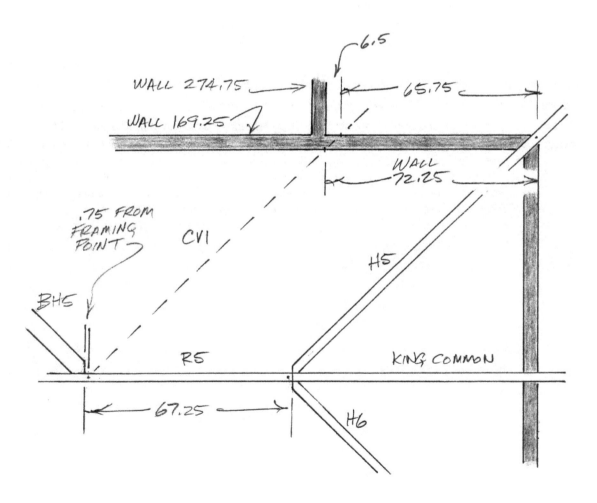

Fig. 2-22b

Question 2-23

Draw a roof skeleton schematic showing the various valley/hip connections applicable for R2 thru R4 including the positioning dimensions.

RCS pg. 75, 127

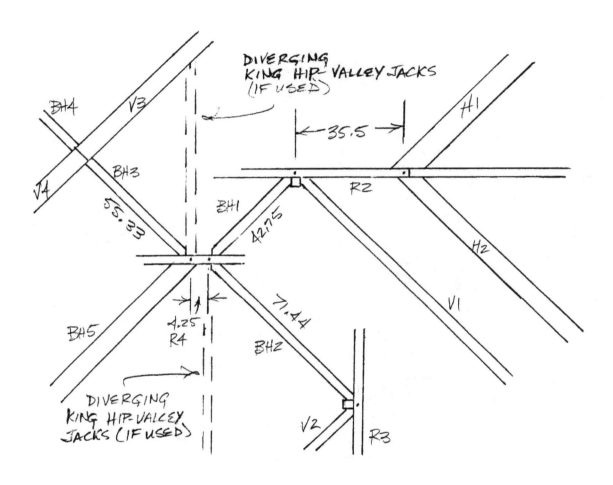

Fig. 2-23

Roof 2 – Phase 2
Monte Sereno CA
6/12 pitch

All ridges, hips, valleys are 2x except as
noted

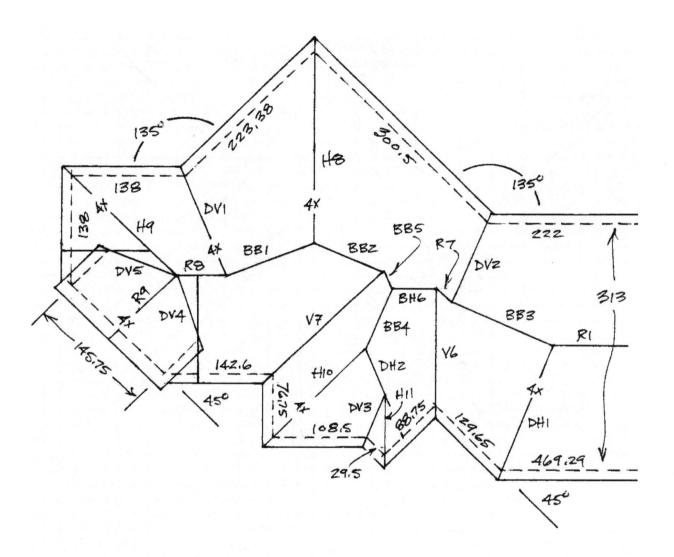

Now on to Phase 2. This part of the roof will require solid concentration. While this part may be intuitive to me now because I know the roof, it wasn't that way at the beginning. When Shone Freeman first sent me the plans I did some real sweating. But it all came down to principles.

Do not let the grand scheme of the plans intimidate you. We will focus on one thing at a time and like a puzzle all the pieces will come together. Resist the temptation to look ahead at my drawing/explanations when you are stumped. Keep trying, for it is the struggle that strengthens you. Do your best to visualize the various roof planes and follow them to their imaginary wall lines. This is the strongest hint I can give you and how I ended up seeing the light.

Phase 2

Vision is the art of seeing things invisible.

Question 2-24

Calculate the common rafters lengths and ridge heights for 2x R8, and dormer 4x R9?

RCS pgs. 50, 120

Answers: 1) R8:

        LL = **133.91**

        TOR = **66.78**

    2) R9:

        LL = **79.52**

        TOR = **66.28**

Math:

    Prelim calcs

        A) *241.06* SOLVED using CM calculator.

            45 Pitch, 145.75 Diag, Run +138, =241.06

    Regular calcs

        1) 241.06 -1.5, ÷2, x1.1180 =**133.91**

           241.06 -1.5, ÷2, x.5000, +6.89 =**66.78**

        2) 145.75 -3.5, ÷2, x1.1180 =**79.52**

           66.78 -.50 =**66.28**

Method:

    A) The span for R8 is found by solving for an imaginary leg in the 45° rt triangle where wall 145.75 is the hypotenuse and adding the result to the length of wall 138. See **Figure 2-24**.

    2) R9 ridge height matches R8 per plan drawings but actual height must be adjusted to accommodate the wider thickness of R9. See Q2-40b.

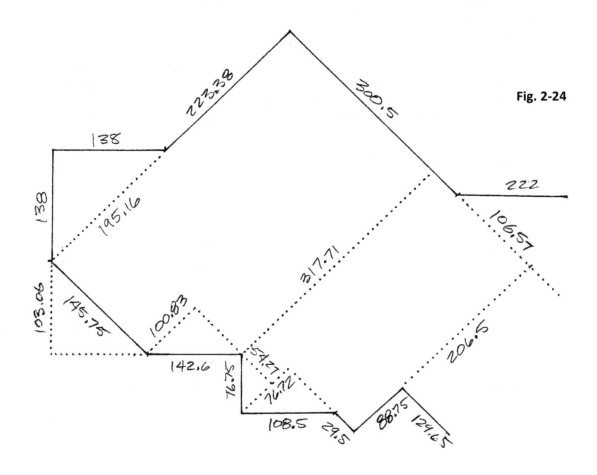

Fig. 2-24

Question 2-25

a) What is the span for R7?

Answer: 206.50

    Math:

      1) SOLVED using CM calculator in steps for clarity. Clear the calculator before each step.

        45 pitch, 138 Run, Diag (195.16)

        142.6 Diag, Run (100.83)

        76.75 Diag, Run (54.27)

        108.5 Diag, Run (76.72)

      2) 223.38 +195.16, -100.83, +54.27, -76.72, -88.75 =**206.50**

    Method:

      1) To find the span for R7 one must calculate the width of the bldg at the inside corner of wall 88.75/wall 129.65 to an imaginary interior continuation of wall 300.5. To do that, start at the 45° corner at wall 223.38/wall 138 and solve all the 45° indentations and projections for legs or hypotenuses of a 45° rt triangle depending on the location.

      2) Then summate all changes (See **Figure 2-24**).

52

b) *What is TOR ridge height of R7?*

Answer: **58.13**

> *Math*: 206.5 -1.5, ÷2, x.5000, +6.89 =**58.14**

Question 2-26

a) *What is the theoretical length of R1?*

> RCS pg. 63, 64, 301

Answer: **247.97**

> *Math*:
>
> > Prelim calcs
> >
> > > *64.82* SOLVED using CM calculator.
> > >
> > > A) 22.5 Pitch, 313 ÷2, Run, Rise (64.82)
> >
> > Regular calcs
> >
> > > 1) 469.29 -(313 ÷2), -64.82 =**247.97**
>
> *Method*:
>
> > A) To find DH1's top offset from 90° to wall 469.29, solve a 22.5° rt triangle for the rise leg when half the bldg span is set as the run leg. See **Figure 2-26a**.
> >
> > 1) Wall 469.29 length less adjustments for H7 and DH1.

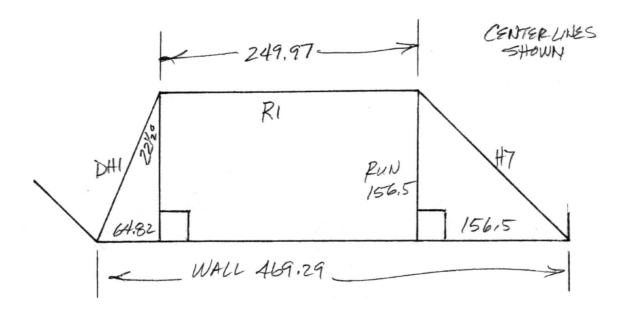

Fig. 2-26a

b) What is the theoretical length of R7?

Answer: **39.45**

> *Math*:
>> Prelim calcs
>>> *42.77* SOLVED using CM calculator.
>>> A) 22.5 Pitch, 206.5 ÷2, Run, Rise (42.77)
>>> Regular calcs
>>> 1) 145.75 +100.83, +54.27, +76.72, +29.5 =407.07
>>> 2) 407.07 -300.5, -42.77, -103.25 =Conv +/- (**39.45**)
>
> *Method*:
>> A) Find DV2's top offset at R7 from 90° to wall 300.5 by solving a 22.5° rt triangle as shown in Q2-26a.
>> 1) Calculate theoretical bldg length from the H8 corner along wall 300.5 to imaginary 90° corner in line with a wall 88.75 interior extension by solving the 45° indentations and projections for legs or hypotenuses of a 45° rt triangle as shown previously.
>> 2) From the theoretical bldg length (407.07) subtract off the length of wall 300.50, the run for V6 (½ span R7 = 103.25) and the top offset distance for DV2 (42.77) to find the ridge length of R7. See **Figure 2-26b**.

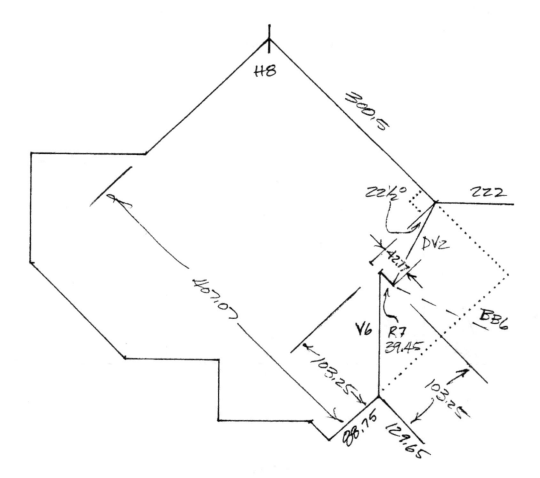

Fig. 2-26b

c) What is the theoretical length of R8?

Answer: **67.40**

  Math:

   Prelim calcs

    49.93 SOLVED using CM calculator.

    A) 22.5 Pitch, 241.06 ÷2, Run, Rise (49.93)

   Regular calcs

    1) 138 -(241.06 ÷2), +49.93 =**67.40**

  Method:

   A) Find DV1's top offset at R8 from 90° to wall 138 by solving a 22.5° rt triangle for the rise leg when ½ the span of R8 is set as the run leg.

   1) From wall 138 length subtract the run for H9 (241.06 ÷2) then add in the top offset of DV1 (49.93) to arrive at the theoretical length of R8. See **Figure 2-26c**.

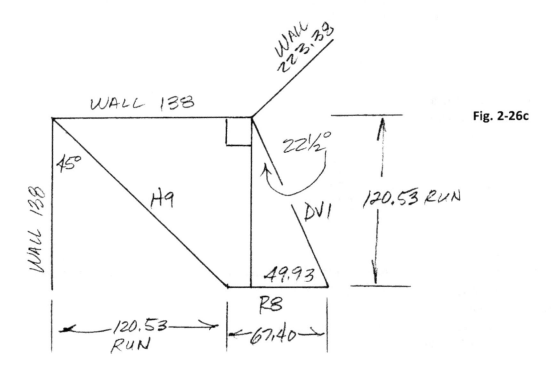

Fig. 2-26c

Question 2-27

What are the lengths of regular rafters V6 and H9 which are run to 2x ridge boards?

RCS pgs. 65, 88

Answers: 1) V6 = **153.75**

    2) H9 = **179.67**

  Math:

   1) 206.5 -1.5, ÷2, x1.5000 =**153.75**

   2) 241.06 -1.5, ÷2, x1.5000 =**179.67**

Question 2-28

a) *Calculate a LL ratio and rafter pitch (x/12) to use for the 45° dogleg hips/valleys on this 6/12 pitch roof.*

RCS pgs. 150-151, 234

Answer: LL ratio **= 1.1923**; rafter pitch = **5.5**/12

Math:

SOLVED using CM calculator.

22.5 Pitch, 12 Run, Diag =Run, 6 Rise, Pitch Pitch Pitch Pitch Conv 8 (**5.5**), Diag ÷12, =**1.1923**

Method:

Use a 22.5° rt triangle and solve for the hypotenuse when the unit of roof run is set as the run leg. Pair this resulting theoretical DH/DV travel together with the unit of roof rise to calculate the pitch and solve for the LL per unit of roof run. Divide this LL by the unit of roof run to calculate the LL ratio.

Note:

This ratio should be saved at the top of the prints for future reference. It will be used many more times before we are done with these plans.

b) *Using this LL ratio, what are the lengths of DV1, DV2 and DH1?*

RCS pg. 150-154

Answers: 1) DV1 = **142.81**

2) DV2 = **122.21**

3) DH1 = **185.70**

Math:

1) 241.06 -1.5, ÷2, x1.1923 =**142.81**
2) 206.5 -1.5, ÷2, x1.1923 =**122.21**
3) 313 -1.5, ÷2, x1.1923 =**185.70**

Question 2-29

a) *What is the bldg's measurement taken perpendicular from wall 300.5 to the inside corner at wall 142.6/wall 76.75?*

Answer: **317.71**

Math: 223.38 +195.16, -100.83 =**317.71**

Method: See **Figure 2-24**

b) *What is the bldg's measurement taken perpendicular from an imaginary interior extension of wall 142.6 to the outside corner at H8?*

Answer: **399**

Math:

SOLVED using CM calculator.

223.38 +195.16, +145.75 <*564.29*> Diag, 45 Pitch, Run (**399**)

Method:

Continue wall 222.38 and wall 142.6 out to form one corner in an imaginary 45° rt triangle as shown in **Figure 2-29b**. Use the length of the hypotenuse incorporating wall 222.38 to solve for the base leg.

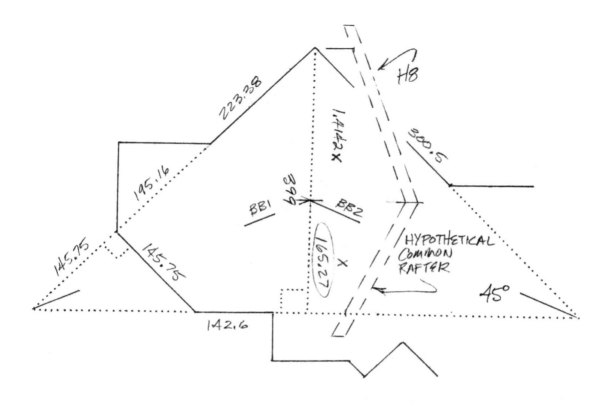

Fig. 2-29b

c) *Using the measurements found in 2-29 a) and b) calculate the theoretical lengths of V7 and H8?*

Answers: 1) V7 = **197.40**

2) H8 = **247.91**

Math:

1) 317.71 ÷2.4142, x1.5000 =**197.40**

2) 399 ÷2.4142, x1.5000 =**247.91**

Method:

2) Divide the measurements from Q2-29a and Q2-29b by the constant 2.4142 (see following question) and multiply the result by the 6/12 H/V LL ratio. See **Figures 2-29b, 2-29c**.

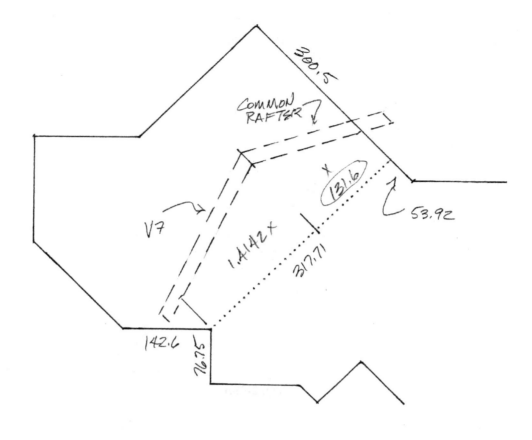

Common Rafter

300.5

V7

1.4142x

x 131.6

53.92

317.71

142.6

76.75

Fig. 2-29c

d) Why is 2.4142 the constant?

Answer: By following in line with V7 or H9 up and across to the far side of the ridge, one would be a descending following a common rafter. Imagine a common rafter butting head-to-head with a hip/valley rafter and covering the 317.71 or 399 distance. This is what is happening in these situations. Remember a hip/valley rafter has a travel of 1.4142 times that of the common rafter run. Therefore, the common rafter run plus the hip travel (1.4142run) equals the given distances (1run + 1.4142run =2.4142run). Dividing these distances by this constant solves for the run.

e) For stacking purposes, what are the theoretical heights at the top framing points of H8 and V7?
   RCS pg. 162

Answers: 1) H8 = **89.53**

2) V7 = **72.69**

Math:

1) 165.27 x.5000, +6.89 =**89.53**

2) 131.60 x.5000, +6.89 =**72.69**

Question 2-30

a) What is the theoretical length and pitch of BB1?

RCS pg. 154-158

Answer: LL = **119.03**; roof pitch = **2.25**/12

Math:

SOLVED using CM calculator.

165.27 -120.53, Run, 6 Inch Pitch, Rise <*22.37*> Stor 1, On/C, Run =Rise, 22.5 Pitch, Diag <*116.91*> =Run, Rcl 1 Rise, Diag (**119.03**), Pitch Pitch Pitch Pitch, Conv 8 (**2.25**)

Method:

Use the dimensional difference between the common rafter runs at each end of the BB (165.27 -120.53 =44.74) as a run measurement and calculate an associated rise (22.37).   Then use the same run measurement again (44.74) as the rise leg of a 22.5° rt triangle and solve for the hypotenuse (116.91). Finally, use this hypotenuse together with the calculated rise (22.38) to compute the LL and pitch.

b) Now make a BB LL ratio one can use to calculate the length of any other BBs in this roof. Make it apply to the difference in run dimensions between the common rafters at each end of a BB (as was done for BB1 in Q2-30a).

RCS pg. 234

Answer: **2.6605**

Math: 119.03 ÷44.74 =**2.6605**

Method:

Take BB1's LL (119.11) and divide it by the difference between the common rafter runs at each end of BB1 (44.74).

Note:

This ratio should be saved at the top of the prints for future reference. It will be used many more times before we are done with these plans.

c) Now, apply this BB LL ratio to solve for the theoretical lengths of BB2 and BB3.

Answers: 1) BB2 = **89.58**

2) BB3 = **141.67**

Math:

1) 165.27 -131.60, x2.6605 =**89.58**

2) 313 -206.5, ÷2, x2.6605 =**141.67**

Question 2-31

a) What is the theoretical length of H10?

RCS pg. 246

Answer: 159.36

Math:

1) 29.5 x1.4142, +108.5 =150.22
2) 12 ÷16.97, x150.22 =106.24
3) 106.24 x1.500 =**159.36**

Method:

1) Solve for the hypotenuse across the wall 29.5 corner (1.4142 x29.5 =41.72) and add that result to wall 108.5 to find it's imaginary extended length if run to intersect wall 88.75
2) Use this dimension (150.22) as the span to set up an unequal roof pitch equation to calculate the run as shown in RCS Figure 12-5.
3) Convert this run into a theoretical hip length using the H/V LL ratio.

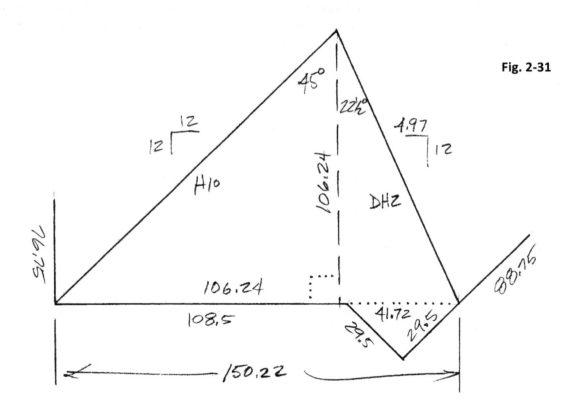

Fig. 2-31

b) For stacking purposes, what is the theoretical height for the top framing point at the intersection of H10 and DH2?

Answer: 60

Math: 106.24 x.5000, +6.89 =**60.01**

Question 2-32

Taking into consideration that DH2 is a supporting dogleg hip and runs full length to wall 88, what is it's theoretical length?

Answer: **126.67**

> *Math*: 106.24 x1.1923 =**126.67**
>
> *Method*: Multiply the run by the earlier calculated DH/DV ratio.

Question 2-33

a) What are the theoretical lengths of DV3 and H11?

Answers: 1) DV3 = **60.04**

2) H11 = **75.54**

> *Math*:
>
> Prelim calcs
>
> A) 41.72 ÷2 =20.86
>
> B) 29.5 +20.86 =50.36
>
> Regular calcs
>
> 1) 50.36 x1.1923 =**60.04**
>
> 2) 50.36 x1.5000 = **75.54**
>
> *Method*:
>
> A) Calculate half the hypotenuse across the wall 29.5 corner.
>
> B) Add the result to wall 29.5 length to find the run.
>
> 1) Multiply the run by the earlier calculated DH/DV ratio to find DV3.
>
> 2) Multiply the run by the H/V LL ratio to find H11.
>
> See **Figure 2-33a**.

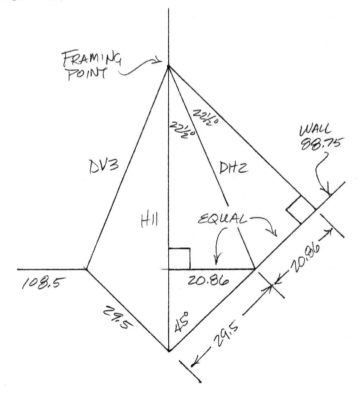

Fig. 2-33a

b) *When stacking, where would DV3 and H11 tie in on supporting DH2?*

Answer: Measure **60.04** uphill on DH2 from the birdsmouth LL mark at wall 88.75.

Method: Use the theoretical LL of DV3 to mark the tie in position at the framing point. See **Figure 2-33b**.

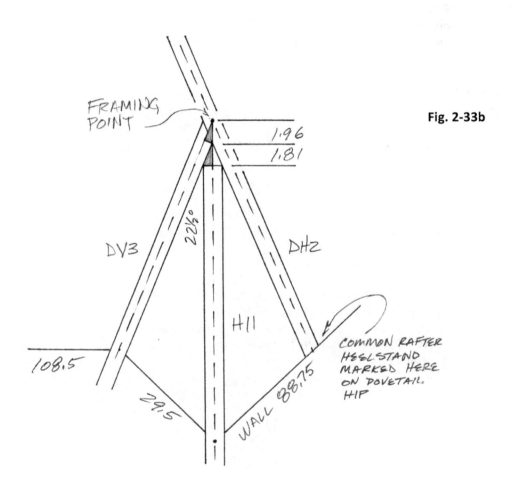

Fig. 2-33b

c) *Draw the DH2 to H11 to DV3 connection. How might one shorten these rafters' headcuts for this connection?*

Answer: 1) Shorten DV3 ½ 45° thk DH2 (1.06) and cut a single 45° cheekcut.

2) Shorten H11 (**3.77**) and square-cut plumb to stack in the joint between BV3 and BH2.

Math:

2) SOLVED using CM calculator.

22.5 Pitch, .75 Rise, Diag <*1.96*> + Run <*1.81*>, =**3.77**

Method:

2) Solve two 22.5° rt triangles for the run leg and hypotenuse when ½ thk 2x is set as the rise leg.

See **Figure 2-33b**.

Question 2-34

a) What is the theoretical length of BH6?

Answer: **55.21**

> *Math*:
> 1) 212.44 +59.25, +206.5 =478.19
> 2) SOLVED using CM calculator.
>   45 Pitch, 478.19 Diag, Run *<338.13>* ÷2.4142 *<140.06>*, -103.25 *<36.81>*, x1.5000 =**55.21**
>
> *Method*:
> 1) Find the length of the hypotenuse in the imaginary 45° rt triangle which incorporates wall 88.75
> 2) Use this length (478.19) to solve for the base leg (338.16). Then divide this number by the constant 2.4142 (explained in Q2-29d), subtract off the run of V6 (206.5 ÷2 =103.25) and multiply result by the H/V LL ratio. See **Figure 2-34a**.

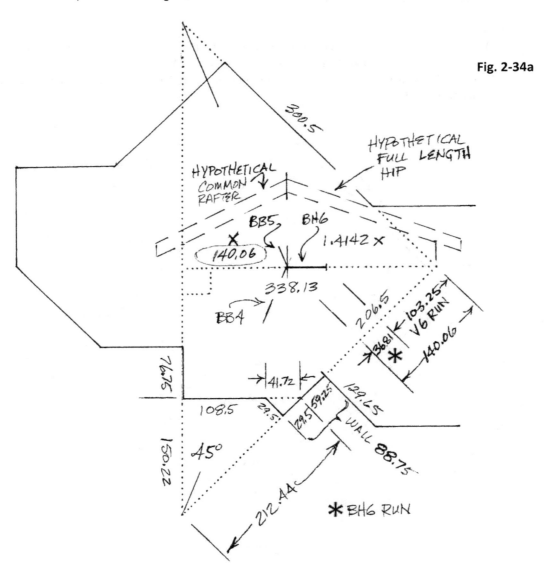

Fig. 2-34a

b) For stacking purposes what is the theoretical height at the top framing point of BH6?

Answer: **76.91**

> *Math:* 140.06 x.5000, +6.89 =**76.92**

Question 2-35

What are the lengths of BB4 and BB5?

RCS pg. 155

Answers: 1) BB4 = **89.98**

2) BB5 = **22.51**

Math:

1) 140.06 -106.24, x2.6605 =**89.98**
2) 140.06 -131.60, x2.6605 =**22.51**

Method:

1) Using the run from wall 300.50 to the high point of BH6 (140.06), subtract off the run of H10 (106.24) and multiply by earlier found ratio for a BB conversion (2.6605).
2) Follow the same procedure described for BB4 but subtract off the run for V7 (131.60) instead of H10. See **Figure 2-35**.

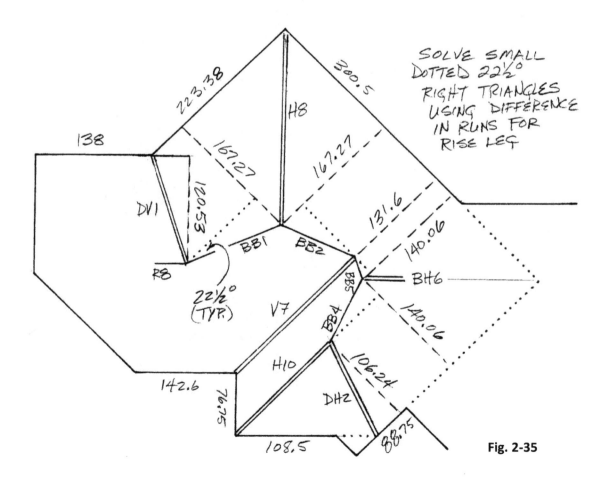

Fig. 2-35

Question 2-36

a) *With the theoretical ridge heights of R9 and R8 being equal, what is the difference in plate heights?*
   RCS pg. 243

Answer: **23.83**

> *Math:* 241.06 -145.75, ÷2, x.5000, =**23.83**
>
> *Method:* R8 span less R9 span divided by 2 and multiplied by the RR ratio.

b) *The side walls for dormer R9 are headers reaching back from the front gable end wall to supporting rafters. How long are these 2 side wall plate beams?*

Answer: **67.40** to SP of 45°

> *Math:* 23.83 ÷.5000, x1.4142 =**67.40**
>
> *Method:*
>
> > Divide the difference in plate heights (23.83) by the RR ratio to find the corresponding run dimension. Multiply this dimension by 1.4142 to convert to a 45° diagonal length. See **Figure 2-36b**.

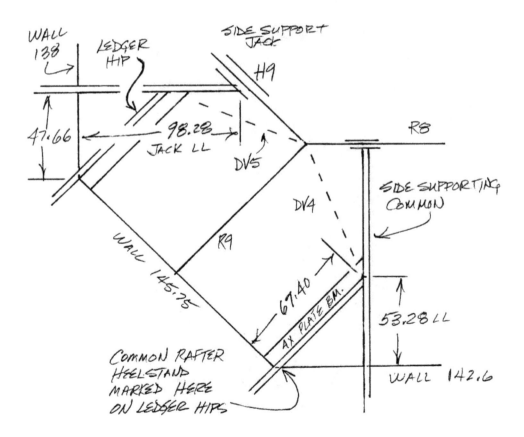

Fig. 2-36b

c) How far from the dogleg bldg corners at each end of wall 145.75 will the dormer's side supporting rafters be positioned?

Answer: **47.66**

        Math: 23.83 ÷.5000 =**47.66**

        Method:

           Convert the difference in plate heights (23.83) into a run dimension by dividing by the RR ratio.

        Note:

           Since the side walls of dormer R9 sit at 45° to the roof plane, they are the hypotenuses in 45° rt triangles formed by the supporting rafters with the outside walls (wall 138 or wall 142.6). Therefore, the run from each dogleg corner to the supporting rafters measured along the outside wall is equal to the run dimension used in determining where on the supporting rafters the side wall plate beams tie in. The run is a function of the rise difference between the two plate heights. See **Figure 2-36b**.

d) Where will the side wall plate-beams tie in to the side supporting rafters?

Answer: **53.28** LL up the supporting rafters from the heelcut LL line.

        Math: 47.66 x1.1180 =**53.28**

        Method: Convert the run dimension found in Q2-36c into a LL measurement.

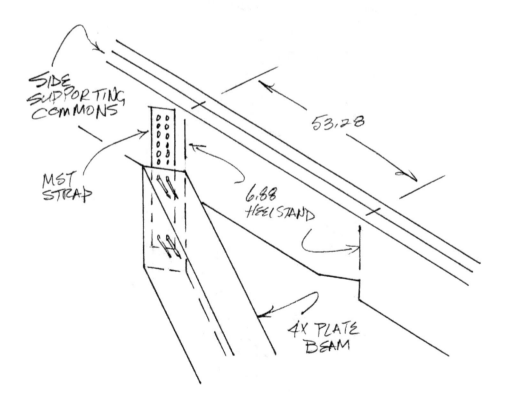

Fig. 2-36d

e) *How might one support the interior end of the side wall plate-beam where it attaches to the side supporting rafters?*
RCS pg. 258 (similar), 40 (similar).

Answer: One method is to use a Simpson MST strap (or similar) run vertically, nailed/screwed to the 45° cut end of the side wall plate-beam which is then nailed/screwed to the supporting rafter using the other side of the strap. Another method is to use a 45° skewed 4x hanger with extended flanges. See **Figure 2-36d**.

Question 2-37

a) *What is the length of the supporting hip jack carrying DV5 and the end of the associated dormer side wall plate beam?*

Answer: **98.24** to LP

Math: 138 -47.66, -2.47 <87.87>, x1.1180 =**98.24**

Method:

From the length of wall 138 subtract off the earlier found supporting rafter bottom position (47.66) and ½ 45° thk H9. The result is multiplied by the COM LL ratio.

b) *During stacking, where would one position the LP of this hip jack on the side of H9?*
RCS pgs. 134-135 (similar)

Answer: Measured on the side of H9, **131.81** from where it crosses the outside wall line.

Math: 138 -47.66, -2.47 <87.87>, x1.5000 =**131.81**

Method: Similar as was done in Q2-37a except multiply 87.87 by the H/V LL ratio.

Question 2-38

What are the lengths and pitch of the wall ledgers along the sides of the dormer at R9?
RCS pgs. 237-239

Answers: 1) **71.49** LL to LP of single 45° cheekcut at head.

2) Pitch = **6/12** H/V

Math: 47.66 x1.5000 =**71.49**

Method:

Since the side walls are at 45° to the roof plane the wall ledgers are hip rafters. Set the heelstand of a common rafter on the side of these rafters facing away from the dormer at that point where they cross the outside wall line. The birdsmouth will have a single cheek 45° heelcut. LL is measured on the top side of the board against the wall. Line up the top corner edge of the SP at the head with the common rafter into which it terminates. See **Figure 2-36b**.

Question 2-39

What are the theoretical lengths of DV4 and DV5?

RCS pgs. 150-154

Answer: **86.89**

> *Math*: 145.75 ÷2, x1.1923 =**86.89**
>
> *Method*:
>
> > Because R9 makes a 45° jog in relation to R8, these rafters are standard dogleg valley rafters. Apply the earlier found ratio to the run.

Question 2-40

a) Draw the top connection at R8/ R9/ H9/ DV4/ DV5. What is the cut-angle at the lower end of DV4 and DV5?

Answer: **67.5°** (See **Figure 2-40a**.)

> *Math*: 90° -22.5° =**67.5°**
>
> *Note*:
>
> > This angle can be made using a swingtable saw in one pass or a regular saw (max bevel adjust 45°) in two passes using the compliment angle cutting method from RCS pgs. 156-157. Another option is to square-cut the rafter plumb at SP similar to Photo 8-6 RCS pg. 154.

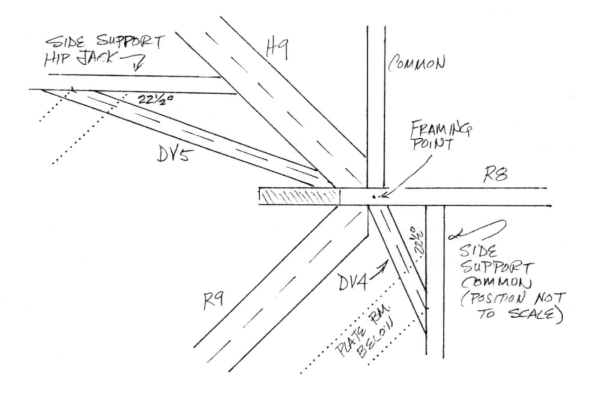

Fig. 2-40a

b) How much lower will 4x R9 connect to 2x R8?

Answer: **.50**

> *Math*: 3.5 -1.5, ÷2, x.5000 =**.50**
>
> *Method:* Multiply ½ difference between the thickness of the two materials by the RR ratio.

c) What is the actual length of R9 not including the outside overhang?

Answer: **97.58**

> *Math*:
>
> > SOLVED using the CM calculator.
> >
> > 145.75 ÷2, Stor 1, 241.06 ÷2, Run, 45 Pitch, Diag <*170.46*>, - Rcl 1, =**97.58**
>
> *Method*:
>
> > Subtract both ½ the R9 dormer span (72.88) and ½ 45° thk of R8 (1.06) from the H9 travel (170.46).
>
> *Note*:
>
> > Imagine if R9 extended to a 90° bldg corner equal in design to the H9 corner (see **Figure 2-24**). If 103.06 is the hypotenuse then half the front face of the dormer (72.88) would be the length of each of the 45° rt triangle legs.

d) What type of cut is made on the ends of R9 and H9?

Answer: Standard dbl 45° cheekcuts. R9 can be cut with a 22.5° cheekcut on the side towards DV4 if so desired.

e) Where are the tips of the dbl 45° cheekcuts on the ends of R9 and H9 positioned along R8?

Answer: .75" towards the cut end of the ridge from the theoretical top framing point. This case is similar to the standard end-of-ridge situation for a regular hip and where 2 king commons butt to each side of the ridge centered on the theoretical top framing point and the hips fit between these and the end of ridge king common.

Question 2-41

a) Draw the R8 to BB1 to DV1 top connection. At what angles should R8 and BB1 be cut where they join? Why?

> RCS pg. 153.

Answers: 1) R8 = **22.5°**, BB1 = **Square-cut**

> 2) One can center the top end of DV1 in the interior corner of the R8 to BB1 connection more easily with these angles since they will match the cheekcuts made at the head of DV1. Installing equal angles of 11.25° on the end of each piece would move the joint off center of a line drawn from the top framing point to the wall 138/ wall 223.38 corner. See **Figure 2-41a**.

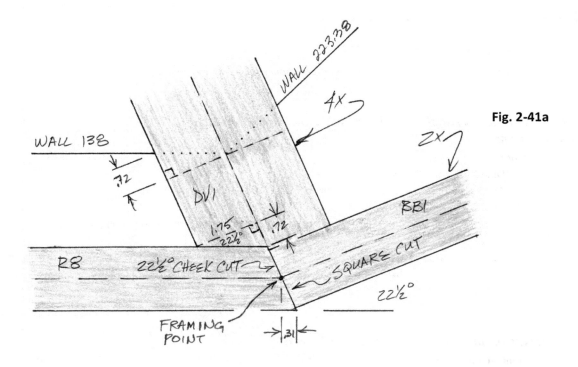

Fig. 2-41a

b) *Does DV1 need to be shortened?*

Answer: No. It was calculated to actual length in Q2-28b.

c) *Draw rafter DV1 showing all the cut-lines. How far measured perpendicular from the top and bottom LL plumb-lines should one position the cut-lines for the various 22½° cheekcuts?*
   RCS pg. 152

Answer: .72

> *Math:*
> SOLVED using CM calculator.
> 22.5 Pitch, 1.75 Run, Rise (**.72**)

Method: Solve a 22½° rt triangle for the rise leg when ½ thk 4x DV1 is set as the run leg.

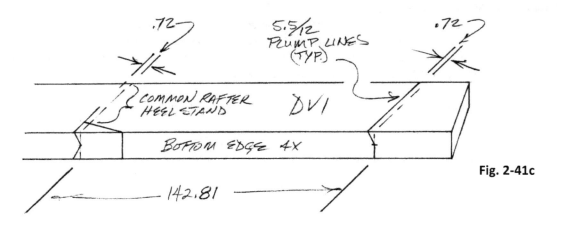

Fig. 2-41c

d) Are there duplicates of this same top connection found elsewhere on the plans?

Answer: Yes, twice more.

1) The R7 to BB3 to DV2 connection is the same except DV1 is a 4x.
2) The R1 to BB3 to DH1 connection is similar but a reverse. While the ridge to BB to DH connection would be cut and assembled similarly, it is an outside corner vs the other being an inside corner. BB3 slopes downhill from DH1 while BB1 slopes uphill from DV1. DH1 is a hip while DV1 is a valley.

Question 2-42

a) Draw the BB1 to BB2 to H8 top connection. At what angles should BB1 and BB2 be cut where they join?

Answer: **22.5°**

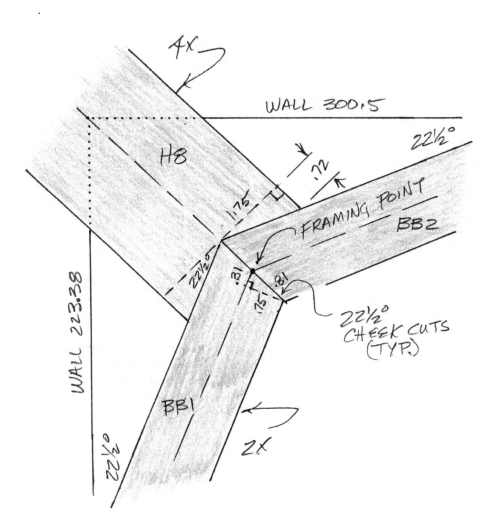

Fig. 2-42a

b) What are the dovetail cheekcut angles on the headcut of H8?

RCS pg. 152 (similar)

Answer: **22.5°**

c) How much is this rafter shortened for the top connection?

Answer: **.81** measured perpendicular to the LL plumb-line at the head.

Math:

SOLVED using CM calculator.

22.5 Pitch, .75 Run, Diag (**.81**)

Method: Solve a 22½° rt triangle for the hypothenuse when ½ thk BB1 or BB2 is set as the run leg.

d) From this newly marked plumb-line for the ridge shortening on 4x H8, how far forward (measured perpendicular) should the cut-lines for the 22.5° dovetail cheekcuts be positioned?

Answer: **.72** (See **Figure 2-42a**)

Math:

SOLVED using CM calculator.

22.5 Pitch, 1.75 Run, Rise (**.72**)

Method: Solve a 22½° rt triangle for the rise leg when ½ thk 4x H8 is set as the run leg.

Question 2-43

How is the BB2 to BB5 to V7 connection similar/dissimilar to the BB1 to BB2 to H8 connection?

Answers: Similar

1) The angles on adjoining ends of BB2 and BB5 are the same 22.5°
2) V7 is shortened the same .81
3) The cheekcut angles at the head of V7 are the same 22.5°

Dissimilar

1) BB1 and BB2 are rising to join at a high point, while BB2 and BB5 are falling to join at a low point.
2) The headcut of V7 is made with a convex dbl 22.5° cheekcut whereas H8 is made with a concave dbl 22.5° cheekcut.
3) The center of V7 planes in with the nearside low point joint of BB2 with BB5, whereas with H8 the top 2 outside edges plane in at 1.75" to each side of the BB1 to BB2 joint (H8 is a 4x).
4) The top cut-lines for H8 measure forward from the shortened plumb-line .72" (4x) while the cut-lines for V7 measure backwards perpendicular from the shortened plumb-line .31" (2x).

Question 2-44

a) Draw the BB5 to BB4 to BH6 top connection. Is it similar to a connection we have already discussed?

Answer: Yes. It is exactly the same as the BB1 to BB2 to H8 connection, except BH6 is a 2x rafter whereas H8 was a 4x.

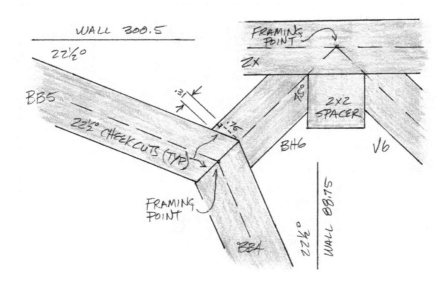

Fig. 2-44a

b) In a normal situation, a BH is calculated to actual LL length via the formula. Due to the irregular situation in these plans, BH6 was calculated to it's theoretical length. Draw this rafter showing top/ bottom shortening and cut-lines. How much is the lower end shortened for the connection at R7?

Answer: **1.06**

Method: ½ 45° thk of R7 (1.06) measured perpendicular to the LL plumb-line.

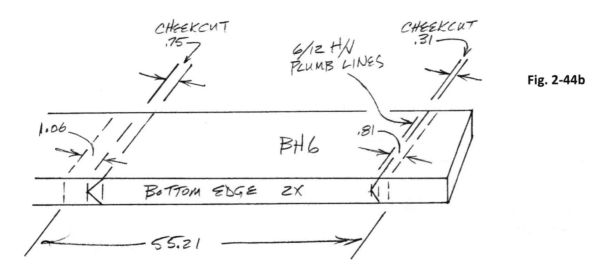

Fig. 2-44b

Question 2-45

a) Draw the DH2 to BB4 to H10 top connection. What is the best way to handle this connection?

Answer: Butt DH2 to BB4 and have H10 coming in from the side.

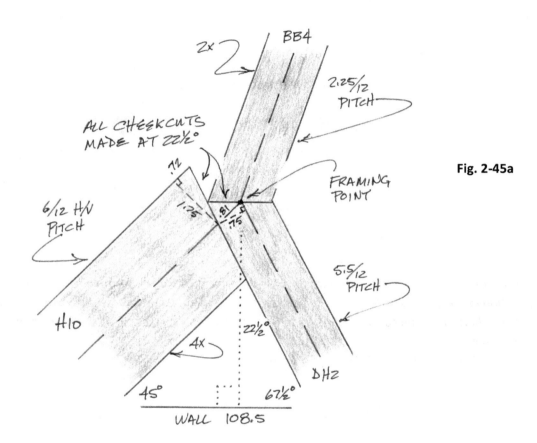

Fig. 2-45a

b) What angles would be installed on each of these rafters?

Answer: All rafters will have 22.5° cheekcuts. H10 is shortened for the connection.

c) How much is H10 shortened?

Answer: **.81**

> *Method*: Solve a 22½° rt triangle for the hypothenuse when ½ thk DH2 is set as the run leg.

74

Question 2-46

How many sets of regular hip/valley jacks are needed in Phase 2 of this roof?

Answer: Hip fill

        full set of 3 -H10 left/H9 rt

        full set of 4 -H9 left/H10 rt

        full set of 6 -H8

        2 full sets of 1 -H11, R9 dormer wall ledgers

      valley jacks

        rt set of 1 -R7

Question 2-47

a) How many sets of regular parallel hip-valley jacks are needed in Phase 2 of this roof?

Answers: 1 jack V7-H10 [2x-4x] = **81.86**

      Math: 76.75 -1.06, -2.47, x1.1180 =**81.86**

b) How many sets of regular diverging hip-valley jacks are needed in Phase 2 of this roof?

Answers: left set of 1 -BH6/V6

Question 2-48

a) What is the 24"OC dogleg hip/valley jack step length?

    RCS pg. 152

Answer: **64.78**

      Math:

        SOLVED using CM calculator.

        22.5 Pitch, 24 Rise, Run, x1.1180 =**64.78**

      Method:

        Solve a 22½° rt triangle for the run leg when the OC spacing is set as the rise leg. Multiply the result by the COM LL ratio.

b) How many sets of dogleg hip/valley jacks are needed in this roof?

Answer: DH jacks

        full set 2 -DH1

        rt set 1 -DH2

      DV jacks

        rt set 2 -DV1

        rt set 1 –DV2

        2 full sets 1 -DV4, DV5 (stackers field cut the top end of the jack from DV5 to H9)

c) For this roof what is the easiest method to cut dogleg hip/valley jacks?

RCS pg. 151-154

Answer: Start with the shortest dogleg jack length being one step measurement to the LP and progressively step up/down from that dimension.

d) To square-cut the jacks plumb at the SP, what would be the length of the shortest jack to it's SP?

Answer: **60.73**

Math:

SOLVED using CM calculator.

22.5 Pitch, 22.5 Rise, Run, x1.1180 =**60.73**

Method:

Solve a 22½° rt triangle for the run leg when the OC spacing less the thk of a jack rafter (22.5) is set as the rise leg. Multiply the result by the COM LL ratio.

Question 2-49

a) What is the 24"OC BB hip jack step length?

RCS pg. 158

Answer: **11.11**

Math:

SOLVED using CM calculator.

22.5 Pitch, 24 Run, Rise <9.94>, x1.1180 =**11.11**

Method:

Solve a 22½° rt triangle for the rise leg when the OC spacing is set as the run leg. Multiply the result by the COM LL ratio.

b) What angle is the single cheekcut at the head of the BB hip jacks?

Answer: **22.5°**

c) What measurement would one use to layout a BB hip at 24" OC if necessary?

Answer: **26.45**

Math:

SOLVED using CM calculator.

22.5 Pitch, 24 Run. Rise <9.94>, x2.6605 =**26.45**

Method: Apply the earlier found BB LL ratio (2.6605) to the run difference for 24" OC BB jacks (9.94).

Book note: Cutting jacks for the remaining unfilled areas of Phase 2 is a messy proposition and requires work. Using the various step measurements previously calculated or combinations of them will help speed things up a bit.

Begin with a scale drawing of the different roof sections and try to fill them in with as many regular jacks of some type (dogleg H/V jacks, diverging hip-valley jacks, etc), then do your best to design areas where you can do short progressions even if it is only a sequence of 2 or 3 jacks. Do not be overly concerned about keeping the OC spacing perfect throughout. As long as the OC spacing is less than or equal to that which is called out on the plans (24" OC in our case) you are fine.

I try to begin a progression with two things in mind: First, it is a point from where I can easily calculate a jack; Second, it is an easily identifiable spot (end of ridge, where a valley rafter crosses the plate, etc.) so those stacking have a reference point of where to begin or place a rafter.

Whatever you end up with, make a good layout map of how the fill was cut for the stackers to follow. And, "of course" mark all the jacks well. I package all the jack fill for a particular section together using 8d nails.

In the questions to follow you may come up with an entirely different layout than I. It may even be better than mine. That's great. What is important is that you know what you did and how you did it. The fastest, simplest way to get the jacks for these messy areas is the goal of this exercise. It will show how well you understood the roof's intricacies.

Question 2-50

a) Sketch the roof section between V6 and DH1. Illustrate how you would cut the jack fill for this section. How many progressions did you find?

Answer: Two progressions.

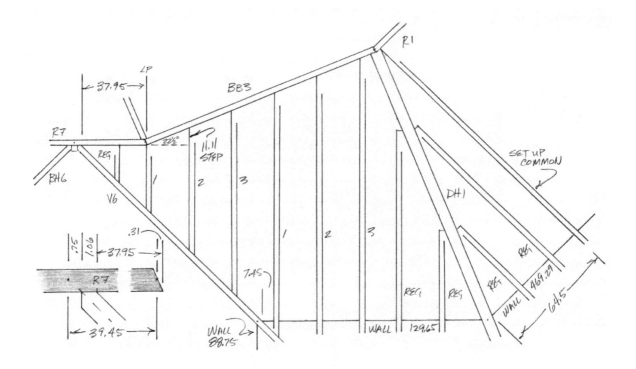

Fig. 2-50a

b) How many regular jacks were utilized in this roof plane?

Answer: One regular valley jack from R7 to V6 and two dogleg hip jack rafters to DH1.

c) What is the step for the BB3 to V6 jacks?

Answer: **37.94**

>*Math*: 26.83 +11.11 =**37.94**
>
>*Method*: Regular hip/valley jack step plus one BB jack step.

d) Why did I choose to start the valley jack progression at the R7/BB3 joint?

Answer: It is the simplest place to start calculation since the theoretical length of R7 is known (39.45). That dimension can be modified into an effective run measurement to start a sequence as is shown in following question.

e) Referring to my jack fill design, what is the length of the first jack in the BB3 to V6 sequence from SP of 22.5° cheekcut at BB3 to SP of 45° cheek at V6?

Answer: **42.43**

>*Math*: 39.45 -.75, -1.06, +.31 <*37.95*>, x1.1180 =**42.43**
>
>*Method*:
>
>Using R7 theoretical ridge length (39.45) subtract both ½ thk ridge and ½ 45° thk valley. Then add the dimension of the 22.5° cheekcut extension at the R7 to BB3 joint (.31 for .75 thk) and multiply the result by the COM LL ratio. See **Figure 2-50a** closeup.

f) How long is the shortest jack in the progression from BB3 to wall 129.65 from heelcut to SP of 22.5° cheekcut at head?

Answer: **147.93**

>*Math*: 206.5 -1.5, ÷2, x1.1180, +(3 x11.11) =**147.93**
>
>*Method*: Common rafter length plus 3 BB jack step lengths.

Question 2-51

a) Sketch the roof section between DH2/H11 and V6. Illustrate how you would cut the jack fill for this section. How many progressions did you find?

Answer: Two progressions.

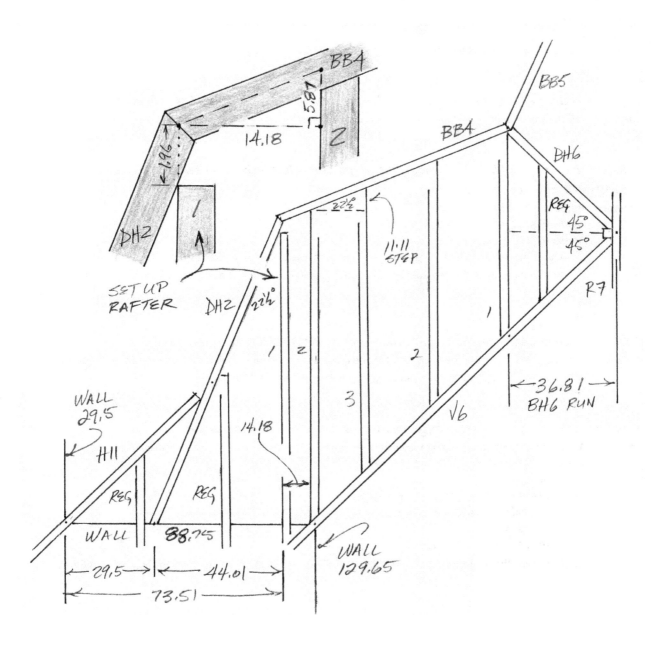

Fig. 2-51a

b) How many regular jacks were utilized in this roof plane?

Answer: One regular diverging hip-valley jack from BH6 to V6, one regular hip jack to H11 and one regular dogleg hip jack to DH2.

c) What is the step for the BB4 to V6 jacks?

Answer: **15.72**

> *Math*: 26.83 -11.11 =**15.72**
>
> *Method*: Regular hip/valley jack step less one BB jack step.

d) Why did I choose to start the BB4 to V6 valley jack progression at the BH6/BB4 connection?

Answer: It is the easiest point from which to begin calculating since the run of BH6 is known (36.81).

e) Referring to my jack fill design, what is the length of this first jack from LP of 22.5° cheekcut at BB4 to SP of 45° cheekcut at V6?

Answer: **80.22**

> *Math*: 36.81 x2, -.81, -1.06, x1.1180 =**80.22**
>
> *Method*:
>
> Double the BH run to find the theoretical length from the BH6/BB4 top framing point to center of V6. Then shorten for ½ 22.5° thk BB4 (.81) and ½ 45° thk V6 (1.06). Multiply the resulting effective run by the COM LL ratio.

f) Why did I choose to begin the second progression at the DH2/BB4 top connection?

Answer: Because the run for DH2 is known (106.24) and can easily be modified. Also, a rafter installed in this position will serve as a "setup" rafter during stacking to help lock in the roof skeleton.

g) What is the length of this setup rafter which runs to the DH2/BB4 connection?

Answer: **116.59** to a plumb square-cut at the head (See **Figure 2-51a**). Or SP of 67.5° single cheekcut if desired.

> *Math*:
>
> Prelim calcs
>
> > *1.96* SOLVED using CM calculator.
> >
> > 22.5 Pitch, .75 Rise, Diag (1.96)
>
> Regular calcs
>
> > 106.24 -1.96, x1.1180 =**116.59**
>
> *Method*: Using the run for DH2 (106.24), subtract ½ 67.5° thk BB4 (1.96) and multiply by the COM LL ratio.

h) How far along wall 88.75 is this rafter located from the H11 corner?

Answer: **73.51** to SP side

> *Math*:
>> Prelim calcs
>>> *44.01* SOLVED using CM calculator.
>>> 22.5 Pitch, 106.24 Run, Rise (44.01)
>>> Regular calcs
>>> 29.5 +44.01 =**73.51**
>
> *Method*:
>> Distance from H11 corner along wall 88.75 to center of DH2 (29.5) plus rise leg of a 22.5° rt triangle (44.01) when the DH2 run distance (106.24) is set as the run leg.

i) Why did I choose to put the second jack in this sequence at the intersection of the outside wall with the side of the valley?

Answer: Locating one end of the rafter at a recognizable point helps during stacking. Here that point is the intersection of the outside wall with the side of the valley. Also, by using the distance along wall 88.75 from the BB4/DH2 setup rafter to the V6 wall corner (88.75 -73.51 =15.24) allows one to rapidly modify the setup rafter's calculated length to create a jack rafter to fit this position as is shown in following question.

j) What is the length of this jack from SP of 22.5° cheekcut at BB4 to LP of 45° cheekcut at V6?

Answer: **124.43**

> *Math*:
>> Prelim calcs
>>> *5.87* SOLVED using CM calculator.
>>> 22.5 Pitch, 15.24 -1.06 Run, Rise (5.87)
>>> Regular calcs
>>> 106.24 -.81, +5.87, x1.1180 =**124.43**
>
> *Method*:
>> Calculate the rise leg of a 22.5° rt triangle when the run leg is set as 14.18 (the distance along wall 88.75 from the setup rafter to V6 (15.24 -1.06 [½ 45° thk V6]). Add this dimension (5.87) to the effective run of the setup rafter (106.24 -.81) and multiply the result by the COM LL ratio.
>
> *Note*:
>> Fit a normal common rafter birdsmouth at the lower end of this rafter. Then cut it tailless with a single 45° cheekcut having the LP follow the heelcut line.

Question 2-52

a) Sketch the roof plane from H10 to DH2/DV3. What special rafters should be cut for this side?

Answer: Only a field cut jack to break the plywood overspan from the longest regular hip jack to DH2/DV3.

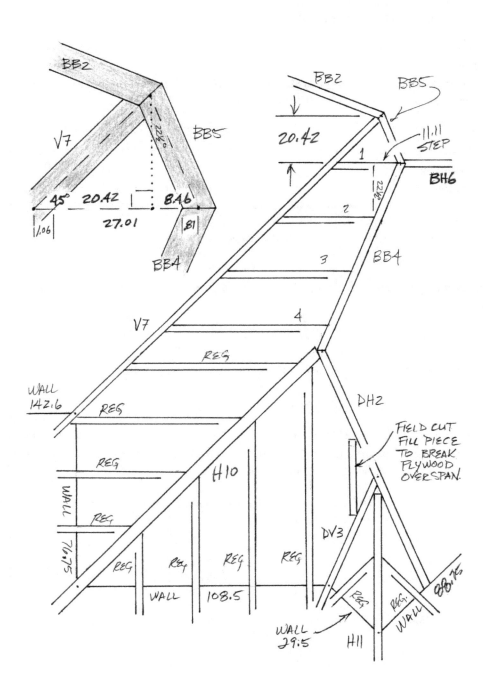

Fig. 2-52a

Question 2-53

a) *Sketch the roof section between V7 and H10. Illustrate how you would cut the jack fill for this section. How many progressions did you find?*

Answer: One progression.   See **Figure 2-52a**.

b) *How many regular jacks were utilized in this roof plane?*

Answer: Three regular hip jack rafters to H10 and one regular parallel hip-valley from H7 to H10.

c) *What is the step for the BB4 to V7 jacks?*

Answer: **15.72**

   Math: 26.83 -11.11 =**15.72**

   Method: Regular hip/valley jack step less one BB jack step.

d) *Why did I choose to start the BB4 to V7 valley jack progression at the BB5/BB4 connection?*

Answer: It is the easiest point from which to begin calculating jacks because the run of BB5 is known (20.42).

e) *Referring to my jack fill design, what is the length of this first jack from LP of 22.5° cheekcut at BB4 to SP of 45° cheekcut at V7?*

Answer: **30.20**

   Math:
   1) 140.06 -131.60 =8.46
   2) *20.42* SOLVED using CM calculator.
      22.5 Pitch, 8.46 Rise, Run (20.42)
   3) 20.42 +8.46 <*28.88*>, -.81, -1.06, x1.1180 =**30.20**

   Method:
   1) Calculate the run of BB5.
   2) Calculate the run for that section of V7 matching BB5's theoretical travel.
   3) Add BB5's run (8.46) together with the matching part of V7's run (20.42). See **Figure 2-52a**. Next, shorten this dimension (28.88) by both the ½ 22.5° thk BB4 (.81) and ½ 45° thk V7 (1.06). Finally multiply the result by the COM LL ratio.

Question 2-54

a) *Sketch the roof section between DV4 and V7. Illustrate how you would cut the jack fill for this section. How many progressions did you find?*

Answer: Three progressions.

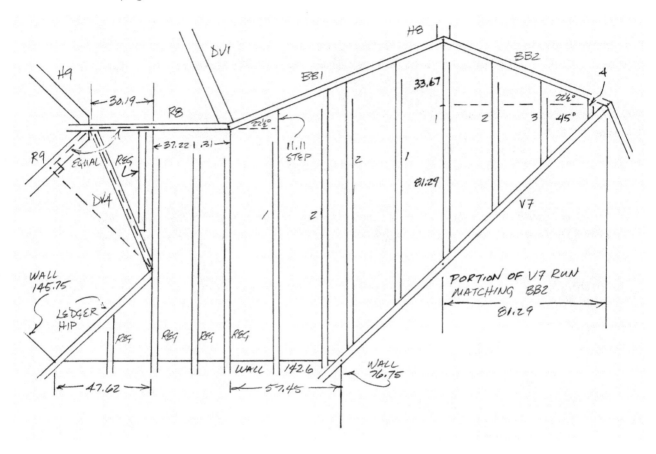

Fig. 2-54a

b) *How many regular rafters were utilized in this roof plane?*

Answer: Four common rafters to R8 (includes one doubler not shown in drawing).

c) *What is the step measurement for the jacks from BB1 to V7? From BB2 to V7?*

Answers: 1) **15.72**

2) **37.94**

Method:

1) Regular hip/valley jack step less one BB jack step.
2) Regular hip/valley jack step plus one BB jack step.

d) Referring to my jack fill design, what is the length of this first jack from LP of 22.5° cheekcut at BB2 to LP of 45° cheekcut at V7?

Answer: **126.43**

Math: 33.67, +81.29, -.81, -1.06, x1.1180 =**126.43**

Method:

Using the same method shown in Q2-53e add the run for BB2 (165.27 -131.6 =33.67) together with that portion of the V7 run matching BB2 (81.29 solved using a 22.5° rt triangle). See **Figure 2-54a**. Next, shorten this dimension by both the ½ 22.5° thk BB2 (.81) and ½ 45° thk V7 (1.06), then multiply the result by the COM LL ratio.

e) What is the length of the first jack (shortest) in the sequence from BB1 to V7?

Answer: **142.15** from SP of 22.5° cheekcut at BB1 to LP of 45° cheekcut at V7

Math: 126.43 +15.72 =**142.15**

Method: Modify the first jack in previous sequence by adding correct step measurement from Q2-54c.

f) What is the length of the first jack (shortest) in the sequence from BB1 to wall 142.6?

Answer: **145** to LP of 22.5° cheekcut at head

Math: 133.88 +11.11 =**145**

Method: Length of R8 common rafter (133.88) plus one BB jack step (11.11).

g) During stacking, where would one mark layout on wall 142.6 for the common rafter positioned at the R8/BB1 connection?

Answer: **85.15** measured from the dogleg bldg corner (wall 145.75/wall 142.6) to the right side of the rafter or **57.45** from the wall 142.6/76.75 corner

Math:

Prelim calcs

A) *30.19* SOLVED using CM calculator.

22.5 Pitch, 145.75 ÷2, Run, Rise (30.19)

B) *.31* SOLVED using CM calculator.

22.5 Pitch, .75 Run, Rise (.31)

Regular calcs

1) 67.41 -30.19 =37.22

2) 37.22 +.31 =37.53

3) 47.62 +37.53 =**85.15** (or 142.6 -85.15 =**57.45**)

Method:

A) Calculate the distance from the end of R8 at H9/DV4 to the dormer side wall supporting commons. This dimension is equal to the top offset found for DV4 at R9.

B) Use a 22.5° rt triangle to solve for rise leg when ½ thk ridge (.75) is set as the run leg.

1) From the R8 theoretical ridge length (67.41), subtract off the ridge section measured from the theoretical top framing point at the R9/H9/DV4 connection over to the dormer side wall supporting commons (30.19).

2) Modify the result (37.22) for the LP of 22.5° cheekcut at the R8/BB1 joint (.31). See **Figure 2-41a**.

3) Summate this dimension (37.53) with the dimension calculated for positioning the dormer's side supporting common rafters from the dogleg corners (47.62).

Question 2-55

a) Sketch the roof section between H8 and DV1. Illustrate how you would cut the jack fill for this section. How many progressions did you find?

Answer: Two progressions.

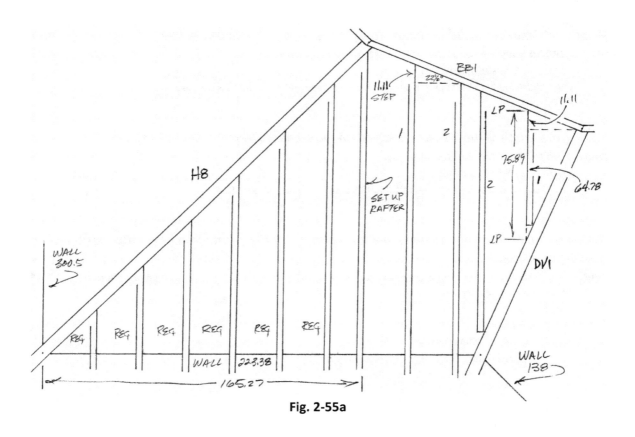

Fig. 2-55a

b) How many regular jacks were utilized in this roof plane?

Answer: Six regular hip jacks to H8

c) What is the step measurement for the jacks from BB1 to wall 223.38? From BB1 to DV1?

Answers: 1) **11.11**

2) **75.89**

Method:

1) BB jack step.
2) BB jack step plus one DV jack step.

d) *What would be the length of a setup rafter run to the BB1/H8 connection?*

Answer: **182.01** to LP of 45° cheekcut at head.

Math: 165.27 -2.47, x1.1180 =**182.01**

Method: H8 run less ½ 45° thk H8 multiplied by the COM LL ratio.

e) *When stacking, where would one position the right side (side with LP of 45° cheekcut) of the set up rafter?*

Answer: At the H8 run distance (165.27) measured along wall 223.38 from the H8 hip corner.

f) *Why do I position this type of setup rafter on one side or the other of the imaginary perpendicular line drawn from the outside wall line to the theoretical top framing point?*

Answer: Positioning it as such simplifies layout and cutting of the rafter because only one cheekcut is made.

g) *When stacking Phase 2 of this roof it is often necessary to preposition a setup rafter at the correct roof pitch to create a height position as a guide for skeleton assembly. Describe how one would use this setup rafter to do this?*

Answer: Use the effective run of the setup rafter to the LP of the 45° cheekcut to calculate a height above the wall plates (rise plus heelstand). Install the rafter "floating in the air" with a vertical brace down to a crossing wall or all the way to the floor which positions the LP's top tip at the calculated height. The hip is then marked with a corresponding attachment point on the side and raised to connect to the prepositioned setup rafter. If this were done with the setup rafter shown in **Figure 2-55a**, the height above the wall plates to the LP's top tip would be (**88.29**) and the distance measured along the side of H8 from it's wall-line crossing to the tie-in connection with the setup rafter's LP is (**244.20**).

Math:
  1) 165.27 -2.47, x.5000, +6.89 =**88.29**
  2) 165.27 -2.47, x1.5000 =**244.20**

Method:
  1) Subtract ½ 45° thk hip from the H8 run, then multiply the result by the RR ratio and add the heelstand of common rafter.
  2) Subtract ½ 45° thk hip from the H8 run and multiply by the H/V LL ratio.

h) *Referring to my jack fill design, what is the length of the first BB jack to the right of the setup rafter, measured from SP of 22.5° cheekcut at BB1 to wall 223.38?*

Answer: **172.76**

Math: 165.27 -.81, x1.1180, -11.11 =**172.76**

Method: run less ½ 22.5° thk BB1 and multiplied by the COM LL ratio. From the result subtract one BB step.

i) What simple method should one use to cut the two BB1-DV1 jacks?

Answer: Cut the shortest as 1 step in length LP-LP (75.89) and the 2nd as 2 steps in length LP-LP (151.78). The headcut at BB1 would be cut with a 22.5° cheekcut while the lower end at DV1 is plumb-cut square at the SP.

j) How does one know where to position these along BB1?

Answer: This is an opportunity to use the earlier calculated 24" OC BB marking step (26.45 from Q2-49c). From the inside intersection of DV1 with BB1 measure along BB1 using multiples of 26.46" to mark where the LP of these jack's 22.5° cheekcut would attach.

Question 2-56

a) Sketch the roof section between DV2 and H8. Illustrate how you would cut the jack fill for this section. How many progressions did you find?

Answer: One progression.

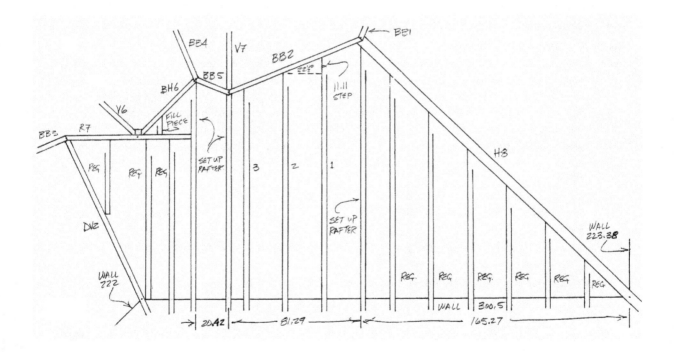

Fig. 2-56a

b) How many regular jack rafters were utilized in this roof plane? How many common rafters?

Answer: Six regular hip jacks to H8 and one regular dogleg valley jack to DV2. Two common rafters are run to an extension of R7.

c) What is the step measurement for the jacks from BB2 to wall 300.50?

Answer: **11.11** (BB jack step)

d) What would be the length of a setup rafter run to the BB2/H8 connection?

Answer: **182.01** (This rafter is a mirror image of the setup rafter run on the other side of H8 to wall 223.38 shown in Q2-55d).

e) Referring to my jack fill design, what is the length of the first jack in the progression to the left of the setup rafter, measured from SP of 22.5° cheekcut at BB1 to wall 300.50?

Answer: **172.76** (This rafter is a mirror image of the longest BB jack in the roof plane on the opposite side of H8 from BB1 to wall 223.38 shown in Q2-55h).

f) What is the length of the setup rafter to the BB5/BB2 connection?

Answer: **146.22** to center of dbl sided 22.5° dovetail headcut.

Math: 131.60 -.81, x1.1180 =**146.22**

Method: V7 run less ½ 22.5° thk of the BB5/BB2 joint multiplied by the COM LL ratio.

g) During stacking, where would one position this setup rafter at the BB5/BB2 intersection along wall 300.50?

Answer: **53.92** from inside corner at DV2

Math: 300.50 -145.75, -100.83 =**53.92**

Method:

The difference between the length of wall 300.50 and the perpendicular distance on the other side of the bldg from an imaginary extension of wall 223.38 to the inside corner at V7 (246.58). See **Figure 2-24** and **Figure 2-29c**.

h) What is the length of the setup rafter to the BH6/BB5 connection?

Answer: **155.40** to LP of 45° cheekcut at head.

Math: 140.06 -1.06, x1.1180 =**155.40**

Method:

Sum of the V6 and BH6 runs (140.06) less ½ 45° thk of BH6 and multiplied by the COM LL ratio.
See **Figure 2-34a**.

i) During stacking, where would one position the setup rafter at the BH6/BB5 connection along wall 300.50?

Answer: **20.42** from center of the BB5/BB2 setup rafter or 33.50 from the inside corner at DV2

Math:

SOLVED using CM calculator.

22.5 Pitch, 140.06 -131.60, Rise, Run (**20.42**)

Method:

Use a 22.5° rt triangle and solve for it's run leg when the difference in rafter runs between V7 and the V6/BH6 combo is set as the rise leg (8.46).

j) In my jack rafter design, why did I continue R7 to tie into the setup rafter run at the BH6/BB2 connection?

Answer: By tying R7 into this setup rafter I accomplished two things. First, by lengthening the ridge it made it easier to stabilize during stacking. Second, the setup rafter will serve as a lateral positioning aid when scribed with the LL of a R7 common (RCS pgs. 125-127).

Question 2-57

a) Sketch the roof section between DV2 and the connection of BB3 with R1. Illustrate how you would cut the jack fill for this section. How many progressions did you find?

Answer: Two progressions.

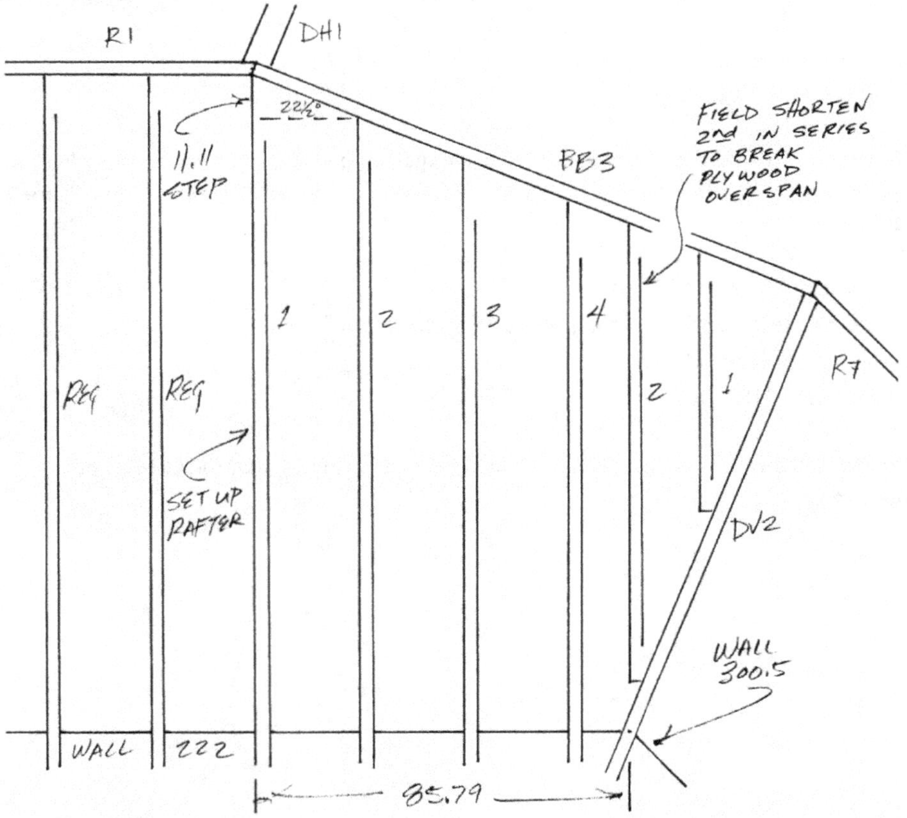

Fig. 2-57a

b) What is the step measurement for the jacks from BB3 to wall 222? From BB3 to DV2?

Answer: **11.11**, **75.89**

> *Method*: See Q2-55c.

c) What would be the length of a setup rafter run to the R1/BB3 connection?

Answer: **174.06** to LP of 22.5° cheekcut at head.

> *Math*: 313 ÷2, -.81, x1.1180 =**174.06**
>
> *Method*: R1 run less ½ 22.5° thk BB3 multiplied by the COM LL ratio.

c) When stacking, where would one position the left side (side with LP of 22.5° cheekcut) of this setup rafter?

Answer: **85.79** from the wall 222/wall 300.5 inside corner.

> *Math*:
>
> SOLVED using CM calculator.
>
> 22.5 Pitch, 103.25 Run, Rise <*42.77*> Stor 1, 156.5 -103.25 <*53.25*>, = Rise, Run - Rcl 1 (**85.79**)
>
> *Method*:
>
> Use the difference in runs between the common rafters for R1 and the common rafters for R7 (156.5 - 103.25 = 53.25) as the rise leg in a 22.5° right triangle and solve for it's run leg (128.56). From this number, subtract off the result found when solving a second 22.5° right triangle for it's rise leg (42.77) while using the run of R7 common (103.25) as the run leg. See **Figure 2-57d**.

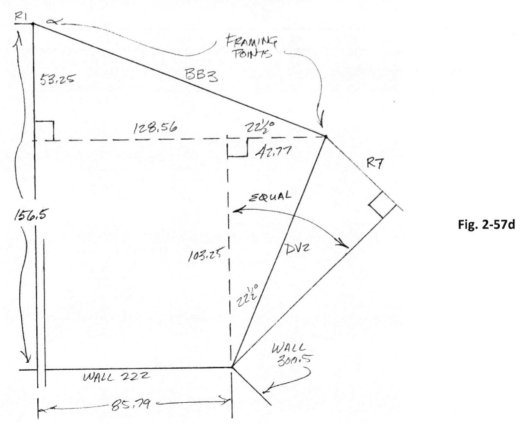

Fig. 2-57d

e) *Referring to my jack fill design, what is the length of the first jack to the right of the setup rafter, measured from LP of 22.5° cheekcut at BB3 to wall 222?*

Answer: **162.95**

> *Math*: 174.06 -11.11 =**162.95**
>
> *Method*: Longest jack less one BB jack step.

h) *What simple method should one use to cut the two diverging BB3-DV2 jacks?*

Answer: As detailed in Q2-55i, cut the shortest jack as 1 step in length from LP-LP (75.89). The 2nd jack would be field cut to break the plywood overspan between the shortest rafter in the jack series from BB3 to wall 222 and the 1st diverging BB3 – DV2 jack. The headcuts at BB3 would be cut with a 22.5° cheekcut while the lower ends at DV2 are plumb-cut square at the SP.

Success is not a place one arrives but rather the spirit which one undertakes and continues the journey.

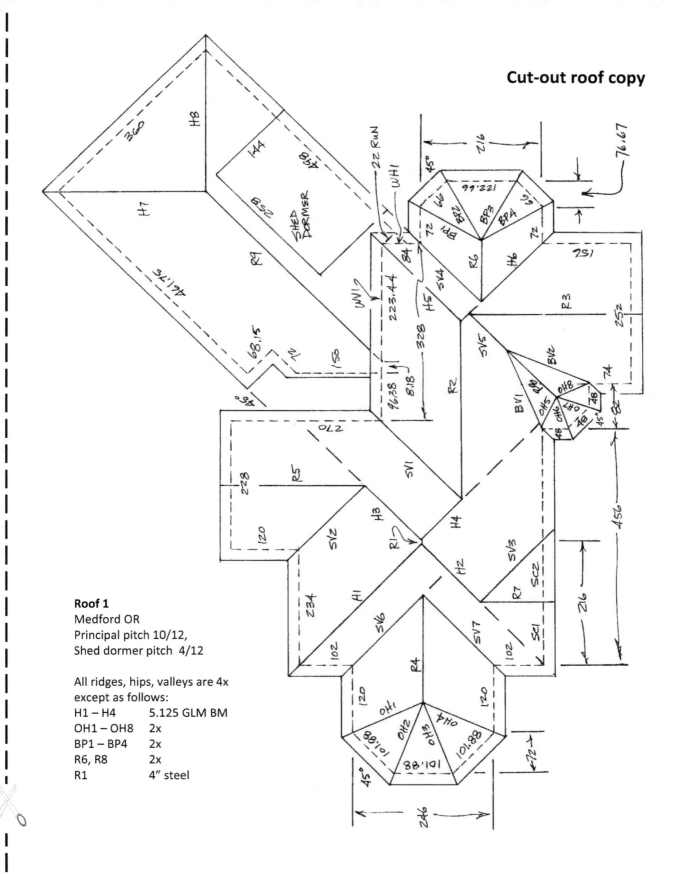

Cut-out roof copy

Roof 1
Medford OR
Principal pitch 10/12,
Shed dormer pitch 4/12

All ridges, hips, valleys are 4x
except as follows:
H1 – H4   5.125 GLM BM
OH1 – OH8   2x
BP1 – BP4   2x
R6, R8   2x
R1   4" steel

Roof 2 – Phase 1
Monte Sereno CA
6/12 pitch

All ridges, hips, valleys are 2x except as noted

Cut-out roof copy

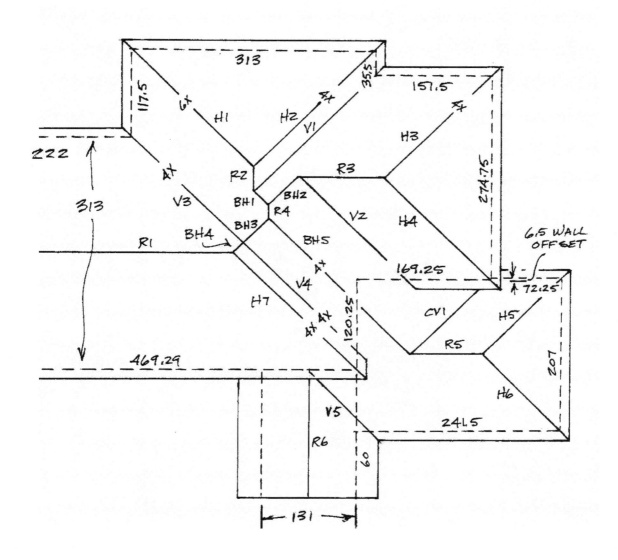

Cut-out roof copy

All ridges, hips, valleys are 2x except as noted

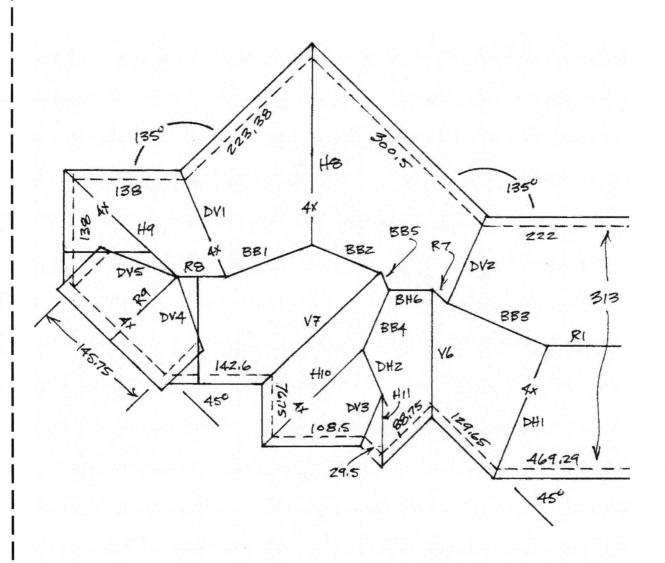

9 780945 186014